IEE MONOGRAPH SERIES 16

Introduction to
diffusion in semiconductors

Titles in the
IEE MONOGRAPH SERIES

1 Techniques of pulse-code modulation in communication networks – G. C. Hartley, F. Ralph, D. J. Tarran and P. Mornet
2 Low-noise microwave amplifiers – H. N. Daglish, J. G. Armstrong, J. C. Walling and C. A. P. Foxell
3 Pal colour television – Boris Townsend
4 Per-unit systems: with special reference to electrical machines – M. R. Harris, P. J. Lawrenson and J. M. Stephenson
5 Layout of e.h.v. substations – R. L. Giles
6 Nuclear power – R. V. Moore (editor)
7 Power diode and thyristor circuits – R. M. Davis
8 Holography and its technology – J. N. Butters
9 Television measurement techniques – L. E. Weaver
10 Power distribution in industrial installations – T. B. Rolls
11 Acoustics for radio and television studios – Christopher Gilford
12 Microwave power measurement – J. A. Lane
13 Random pulse trains: their measurement and statistical properties – C. H. Vincent
14 Electrical insulating materials and their application – R. W. Sillars

IEE MONOGRAPH SERIES 16

Introduction to
diffusion in semiconductors

Brian Tuck, B.Sc., Ph.D., M.Inst.P.
Lecturer,
Department of Electrical & Electronic Engineering,
University of Nottingham,
Nottingham, England

PETER PEREGRINUS LTD.
on behalf of the
Institution of Electrical Engineers

Published by Peter Peregrinus Ltd.,
Southgate House, Stevenage, Herts. SG1 1HQ, England

First published 1974

ISBN: 0 901223 56 5

Typeset by Santype Ltd. (Coldtype Division), Salisbury, England
Printed in England by W & J Mackay Limited, Chatham by Photo-Litho

Foreword

Diffusion is one of the most important processes employed in the semiconductor industry. It is used at least once in the manufacture of most devices. In the past, much diffusion work has been carried out on an empirical basis, but with the increasing sophistication of semiconductor devices, such an approach is becoming quite inadequate. A great deal of research has been carried out in the last decade and our understanding of the subject has greatly increased. This is especially true of diffusion in compound semiconductors, many of which have become available in high-purity single-crystal form only in recent years.

This book is an attempt to give a fairly simple account of the basic principles of diffusion in both elemental and compound semiconductors. I hope it will be understood by the wide range of scientists and engineers involved in this type of work, and also by students studying semiconductors at senior undergraduate or postgraduate level. Throughout the text examples are taken from the literature to illustrate the points made. If it appears that silicon is used in a disproportionate number of examples for elemental semiconductors, then my defence is that this is a true reflection of the state of the literature. Similarly, for compounds, gallium arsenide is so much better understood than any other compound semiconductor that this material also appears many times in the examples given in the book. Indeed Sections 3.17 and 6.9 taken together represent a fairly complete description of our understanding of the zinc/gallium arsenide diffusion system as it stands at present.

When I started working in this field, I found that my main troubles were concerned not so much with the difficulty of the subject, which is no more or less intellectually demanding than other subjects, but with its interdisciplinary nature. Things had to be understood that were outside the discipline I had been nurtured in and, much worse, that were written in entirely different brands of technical English. It is incredible how differently a chemist and an electrical engineer, for instance, can express the same idea. The book is an attempt to spare the reader some of that frustration, and present as simply as possible the elements of the subject written in a single brand of English.

At the end of the book there are two appendixes containing diffusion data and also warnings about using it incorrectly.

Contents

Principal symbols used in text

A	Avagadro's number
A	an acceptor
a	activity coefficient
B, B'	mobility (There are two common definitions of mobility, one used in electrical work when we are usually concerned with electron and hole mobilities, and one used by chemists. Both are used in the text; the definitions are given in Section 2.2.8)
b	Burgers' vector
C	concentration of atoms
C	number of components (in phase rule)
D	diffusion coefficient
D	a donor
E	energy level in semiconductor
$\mathscr{E}$	electric field
e	charge of electron
e	an electron (in a chemical reaction)
F	Helmholtz free energy
F	number of degrees of freedom (in phase rule)
$\mathscr{F}$	force acting on a particle
F	Fermi energy (temporarily, in Chapter 6)
f	correlation factor
G	surface conductance
G	Gibbs free energy
g	free energy of 1 mole of pure liquid
H	enthalpy
h	a hole
h	Planck's constant
I	current
I	an interstitial atom
J_x	flux of atoms in x-direction (similarly y, z directions)
J, j	current density
K	equilibrium constant
k	Boltzman's constant
k	segregation coefficient
M	amount of diffusing material
M_{As}	atomic weight (of arsenic, in this case)
m_e, m_h	mass of an electron, hole
N_1	mole fraction of component 1
n	electron concentration
P	pressure
P	number of phases (in phase rule)

P_v	probability of a site containing a vacancy
$\mathscr{P}$	concentration of ion pairs
p	concentration of holes
Q	amount of heat, activation energy
R	gas constant
R	electrical resistance
r	jump vector
S	entropy
S	a substitutional atom
T	temperature
t	time
U	internal energy
V	volume
V	vacancy
V	volts
v	velocity
W	weight
$\mathscr{W}$	number of microscopic states
Y	Young's modulus
Z	number of nearest neighbour sites in a crystal
Z_j	number of states in the jth energy level
z	number of electronic charges on an ion
α	amount of diffusing material
α	jump distance
β	lattice contraction coefficient
γ	activity coefficient
ϵ	permittivity
μ	chemical potential
ν	Poisson's ratio
ν, ω	jump frequencies
ρ	dislocation density
σ	conductivity
σ	stress
Ω	mass-action constant for ion pairing reaction

Chapter 1

Diffusion and semiconductors

Diffusion is one of the most familiar of natural phenomena. If a small amount of coloured liquid is introduced at one end of a column of water, the sharp division between the dye and the water is gradually obscured, and after a short period the column is uniformly coloured. The dye is said to have diffused into the water and it is a simple matter to show that the effect has nothing to do with mechanical agitation or convection. The experiment is a demonstration of the fact that in a system in which particles can move in a random manner, the net result of the movements will be to iron out concentration gradients. The phenomenon is important in all states of matter; this book is concerned with the technologically important group of solids called semiconductors.

Strictly, the existence of a concentration gradient is not in itself a sufficient reason for atoms to change their distribution. If a large enough force is applied to the atoms, they might be induced to move in such a way as to increase the concentration gradient, and examples of 'up-hill' diffusion do occur. The most general statement of the phenomenon is that particles will move if there is a gradient in chemical potential. However, in most practical cases the gradient in chemical potential is caused by the concentration gradient, and it is this latter quantity that will be considered for most of the book.

The mathematical basis of the subject was laid by Fick, whose first law sets forward the hypothesis that the rate of transfer of diffusing particles through unit area is proportional to the magnitude of the gradient measured normal to the area. Thus if the x axis is chosen as the direction of a concentration gradient, the flux of atoms in the x direction J_x, is given by

$$J_x = -D\frac{\partial C(x,t)}{\partial x} \tag{1.1}$$

where $C(x, t)$ is the concentration of atoms, which depends on both time and distance, and D is the constant of proportionality, usually called the diffusion coefficient. If $C(x, t)$ is expressed in terms of a number of atoms per cubic metre, i.e. m^{-3}, then the units of J_x are $m^{-2}\ s^{-1}$ and those of D are $m^2\ s^{-1}$. The negative sign on the right hand side of eqn. 1.1 indicates that diffusion takes place down the diffusion gradient, from high to low values of C. The equation is, of course, no more than a hypothesis until checked by experiment. In one important respect, however, it agrees with common observation; when the concentration gradient goes to zero, so does the flow of matter. It is worth noting the similarity between eqn. 1.1 and Fourier's law of heat conduction, which states that the flow of heat is proportional to the temperature gradient, and also Ohm's law, which predicts proportionality between the flux of current and the potential gradient.

The equation may be written more generally expressing flux as a vector, **J**,

$$\mathbf{J} = -D\, \nabla \mathbf{C} \tag{1.2}$$

This is correct only in an isotropic medium, however, in which the value of D is the same in all directions. In anisotropic mediums, the vectors **J** and $\nabla \mathbf{C}$ are not, in general, parallel and the problem becomes more difficult. Anisotropy will be considered in the following chapter.

Diffusion occurs as a result of the random motion of particles and these are always thermally activated. The diffusion coefficient is therefore a very strong function of temperature, T and a relation of the form

$$D = D_o \exp -\frac{Q}{kT} \tag{1.3}$$

is often followed, where D_0 is constant and Q is some activation energy. Since almost all heat treatments take place at constant temperature in practice, a diffusion coefficient obeying a relation of the type 1.3 is constant for a given experiment. The expression 'the diffusion coefficient is a constant', which is habitually used in this subject should therefore be taken to mean that it is a constant at constant temperature. This temperature dependence is very useful from the point of view of the technology of materials. It is common in metallurgical processes, for instance, to achieve a certain required configuration of the atoms of a material at high temperature, and then to reduce the temperature as quickly as possible. This process, known as quenching, freezes in the high-temperature situation, which is unlikely to be the correct equilibrium configuration at room temperature. The atoms know this quite well, of course, and proceed to diffuse to their correct positions in the crystal. The values of diffusion coefficients in solids at room temperature are so low, however, that the achievement of equilibrium might take some thousands of years, which is usually quite long enough for the state to be considered stable. Even a slow cooling of a solid is usually not slow enough to maintain thermodynamic equilibrium down to room temperature. The condition of most solids is therefore more representative of the last high temperature process undergone than of anything else. This is especially true of naturally-occurring solids, such as rocks.

Elemental semiconductors come from group IV of the periodic table; the most important are silicon and germanium, and these are the most extensively investigated semiconductors. A wide range of compounds also act as semiconductors. The most complete diffusion studies have been carried out for those compounds made between atoms of groups III and V, and groups II and VI of the periodic table. The useful semiconducting III–V compounds consist of the various combinations between In, Ga and Al on the one hand and P, As and Sb on the other. There are also solid solutions between compounds, such as $Ga_{(1-x)}Al_xAs$ and $GaAs_{(1-x)}P_x$, where $0 \leqslant x \leqslant 1$. The II–VI compounds have the distinction of having been investigated for a longer period of time than other compound semiconductors; they were studied long before the word 'semiconductor' was coined. The most important of these are the compounds formed between Cd and Zn from group II and S and Se from group VI. A number of more exotic materials also show semiconducting properties. These are semiconductors of the 'glassy', 'amorphous' and 'organic' types, and will not be considered in this book. Almost all applications of semiconductors use single-crystal specimens and this will be assumed here, unless otherwise stated.

The special electrical properties of semiconductors arise very largely from the fact that they have two mobile species which can carry an electric current; negative electrons and positive holes. There are therefore three forms that a given semiconductor can assume; intrinsic, in which the numbers of electrons and holes are equal, *n*-type, in which the number of electrons exceeds the number of holes, and *p*-type, in which holes predominate. A completely pure semiconductor is normally intrinsic, and the other two types are formed by adding impurities or 'doping' atoms. The addition of dopant is normally carried out either by adding the material during the original crystal growth, or by diffusing it in from an external source at some later stage. In the case of the group IV semiconductors, *p*-type material is usually formed by adding atoms from group III (acceptors), *n*-type by using group V (donors). Very small concentrations of added impurity can bring about large changes in electrical characteristics. Taking germanium as an example, the conductivity of intrinsic material is $2{\cdot}3(\Omega\text{m})^{-1}$. If a donor-type impurity is added to the extent of only one atom to every 10^8 germanium atoms, the conductivity increases to $26{\cdot}9(\Omega\text{m})^{-1}$. This gives some idea of the order of purity that must be achieved in semiconductor work. There is little point in carefully adding one atom in 10^8 of some doping material, if there is already an impurity level greater than this in the crystal, by accident.

By controlling the doping levels in a semiconductor crystal, it is therefore possible to control both the magnitude of the conductivity, and the type of conductivity. Modern semiconductor devices rely on getting different amounts of *n*-type and *p*-type dopants in different parts of the same crystal by a combination of diffusion and masking procedures. For instance, a resistor in an integrated circuit might be made by introducing a strip of medium-conductivity *n*-type material into a substrate of high-resistivity silicon. Similarly, a bipolar transistor is a *p–n–p* or *n–p–n* sandwich and, in general, the *p–n* junction is the building block for a large number of devices. It is a useful exercise to consider in general terms what must be done to make such a junction. To fix ideas, assume that a *p–n* junction is to be made in a semiconductor slice which has been doped *n*-type during growth, with concentration $C_D \mathrm{m}^{-3}$. The junction is shown in Fig. 1.1*a*. It is obvious that a *p*-type dopant must be added to the crystal, and this can be achieved by introducing some phase containing the acceptor impurity adjacent to the upper face of the crystal slice at some elevated temperature. This might be done either by covering the top surface with a solid film containing the acceptor, or by surrounding the slice with a vapour. In the latter case it would be necessary to cover the sides and bottom face of the slice with some masking layer, so that acceptors entered only through the top face.

The semiconductor will then begin to come to equilibrium with the external phase, starting with the top surface. A concentration of acceptors $C_A \mathrm{m}^{-3}$, will quickly be achieved at the surface, where C_A is the solubility of the acceptor in the semiconductor under the conditions of the experiment. If $C_A > C_D$, then this means that the surface becomes *p*-type. A concentration gradient now exists within the solid, and the acceptors start to diffuse into the slice. After a time, *t*, a concentration profile exists as shown in Fig. 1.1*b*. That region close to the surface, for which $C_A > C_D$ is *p*-type, and the *p–n* junction occurs at the depth for which $C_A = C_D$. As *t* increases so does the junction depth. If the experiment were continued indefinitely, the diffusion process would cause equilibrium to be achieved and the sample would become homogeneous, with no *p–n* junction. It follows that it is rather important to know when to stop the process by quenching the slice down to room temperature.

It is apparent that the two properties of diffusion and solubility are equally important and they are, in fact, often investigated together. The concentration of diffusing material at a point inside the semiconductor depends not only on the diffusion coefficient, but also on the value of C_A at the surface. If in the example

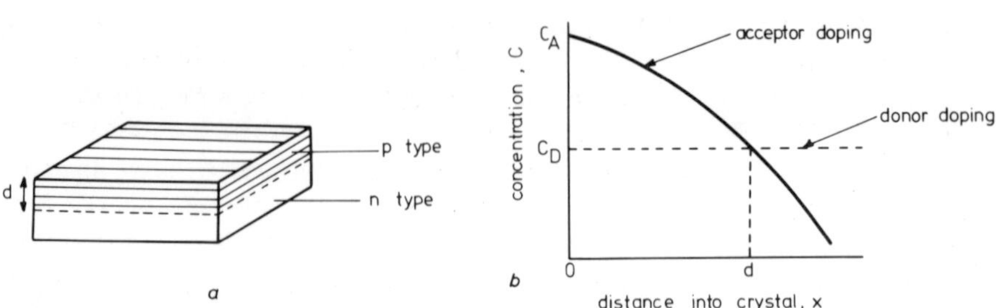

Fig. 1.1 *a* *p–n* junction made by diffusing an acceptor from the top surface

b Concentration profile of the acceptor, showing the depth of the junction

the solubility proved to be less than C_D, then it would not be possible to make a p–n junction at all. Solubility is temperature dependent for a given system, and also depends on the concentration of the dopant in the external phase. When the external phase is simply a vapour, this amounts to saying that C_A depends on vapour pressure. The ratio between the concentration of the doping atom in the external phase, and the equilibrium concentration in the semiconductor, is called the segregation coefficient. It is often, but not always, a quantity which depends only on temperature.

Surrounding a semiconductor homogeneously doped with an acceptor there should therefore be the 'correct' vapour pressure (there is also a 'correct' vapour pressure of the semiconductor material itself, of course). If the pressure is greater than this correct (or equilibrium) pressure, then solute atoms will leave the vapour phase to join the crystal lattice. If the pressure is less than the equilibrium pressure, the dopant will leave the solid phase. During a diffusion the specimen is not homogeneous, but the assumption is usually made that the surface is in equilibrium with a surrounding vapour. Returning to the example of Fig. 1, these considerations suggest an interesting situation with regard to the original donor doping, C_D. In the experiment described, no external vapour pressure of the donor material was provided. Whatever the equilibrium vapour pressure for the donor material is, it is unlikely to be zero, and so it is to be expected that donors would diffuse out of the semiconductor as the acceptors diffuse in. This situation probably occurs more frequently than is generally realised.

Apart from the large dependence on temperature, the diffusion coefficient can vary with other parameters in various systems. Dependence of D on the concentration of the diffusing species is quite common, and it can also depend on time, concentrations of other dopants and defects in the crystal etc. As far as diffusion in semiconductors is concerned, it is true to say that a 'simple' diffusion coefficient is the exception rather than the rule. The quantity D_0 in eqn. 1.3 is therefore not usually a simple constant for a semiconductor system.

Vacancy concentration usually plays a major role in diffusion in solids. Any crystal contains a certain number of vacancies; at temperatures close to the melting point the number of vacant lattice sites might typically be of the order of 1 in 10^4. For a pure elemental semiconductor the vacancy concentration at thermodynamic equilibrium is a function of temperature. To say that the diffusion coefficient depends on vacancy concentration under these circumstances is merely to repeat that it is temperature dependant. However under practical conditions, one must consider the likely state of the semiconductor at the start of the experiment. Presumably the last time it was at equilibrium was during crystal growth: it would then have had the (rather large) concentration of vacancies appropriate to a temperature just below the melting point. The crystal would then have been subjected to the traumatic experience of fast cooling down to room temperature. During this process it would have found itself in the position of having too many vacancies, with very little time to do anything about it. For the period during which the crystal was still at a high temperature, the number of vacancies could have been reduced by diffusing to the surface. A second mechanism that can be resorted to by vacancies deeper within the crystal, is for the vacancies to congregate together in flat plates in order to reduce energy. The plates can collapse giving rise to ring dislocations, but eliminating the vacancies. Any excess vacancies not eliminated by either of these means would have been frozen into the crystal, so that it is probable that at room temperature the elemental semiconductor would have more vacancies than it should.

When a diffusion experiment is carried out on the crystal, it is again raised to a temperature close to its melting point, and it sets about achieving the high number of vacancies appropriate to that temperature. The process will take time, however, and if this time is not very short compared to the duration of the diffusion experiment, then it will affect the results.

The situation with respect to compound semiconductors is very much more complicated. There is now more than one kind of vacancy and the crystal is allowed some variation in stoichiometry at a given temperature. Specifying the temperature no longer determines the vacancy concentrations at equilibrium; the vapour pressure of one of the elements in the vicinity of the semiconductor must also be stated. Consider, for example, the compound GaAs. The numbers of gallium and arsenic vacancies in the crystal are fixed by determining the temperature, and either the gallium or the arsenic vapour pressure, in the surrounding atmosphere; let us choose the latter. The vacancy concentrations can be changed by altering the arsenic overpressure and keeping the temperature constant. It works out that at constant temperature the product of the two concentrations must be constant to maintain equilibrium. There is therefore a good deal of scope for altering the vacancies without changing the temperature, and the problems described above for elemental semiconductors have an extra dimension in compounds. By the time the GaAs crystal has been grown, quenched to room temperature and raised to diffusion temperature, it is most unlikely that the vacancy concentrations bear any resemblance to the equilibrium concentrations at the diffusion temperature and arsenic overpressure chosen by the experimenter. It may well be that the experimenter is under the impression that in his experiment all that is happening is that zinc, say, is being diffused into GaAs. In fact, the 'correct' vacancy equilibrium is also being approached (also by diffusion, from the surface and from dislocations) at the same time. This latter effect could easily be the major diffusion effect that is occurring.

The fact that changing stoichiometry can have an effect on a diffusion experiment is illustrated in the following (Tuck, 1969)*: Fig. 1.2 shows two radio-tracer diffusion profiles for zinc diffusing in GaAs. The experimental conditions for the two specimens were identical; both diffusions were carried out in an atmosphere of $4{\cdot}1 \times 10^3$ Nm^{-2} of arsenic, with sufficient zinc in the ambient

* Tuck, B.: *J. Phys. & Chem. Solids*, 1969, **30**, p. 253.

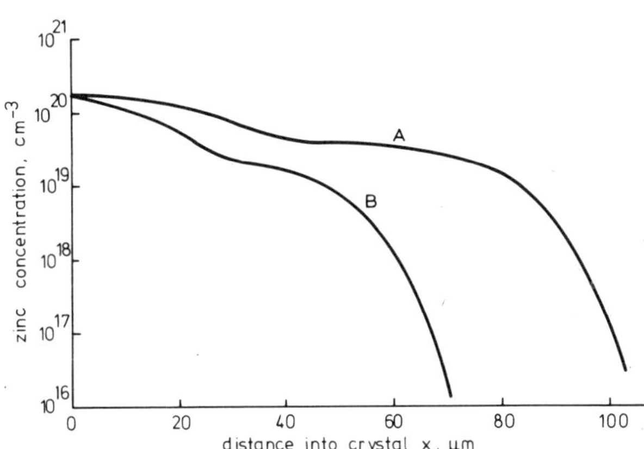

Fig. 1.2 Diffusion profile for zinc in gallium arsenide at 1000°C. For profile A stoichiometry changes were taking place during diffusion, but for profile B only diffusion was occurring

vapour phase to give a surface concentration of $1{\cdot}5 \times 10^{26}$ m^{-3}. The temperature of diffusion was 1000°C. Before taking profile B, however, the semiconductor slice was annealed at 1000°C for about two weeks in an arsenic atmosphere of $4{\cdot}1 \times 10^3$ Nm^{-1}, for it to achieve the appropriate stoichiometry. It was quickly quenched before being reused in the diffusion experiment. Profile A was taken without these preliminaries. The difference between the profiles is apparent and far in excess of normal experimental variation. It seems reasonable to assume that the difference is due to the fact that in experiment A both diffusion and stoichiometry changes were taking place, whereas in B only diffusion was occurring.

This qualitative account, given to acquaint the reader with the type of consideration that is important in diffusion, is still rather oversimplified. No mention has been made of possible interactions between diffusing species and vacancies or interstitial atoms, of reactions in the surrounding vapour, or of liquid phases being formed. All of these points will be taken up in later chapters.

Chapter 2

Solutions to diffusion equation

2.1 Fick's second law

The second law is derived by applying a continuity condition to eqn. 1.1. Consider the flow of diffusing material through an element of volume $\Delta x \Delta y \Delta z$, as shown in Fig. 2.1. There are components of matter flow in all three directions, the y and z flows are omitted from the figure for the sake of clarity. The continuity argument is simply that the rate at which material is accumulating in the element is equal to the rate at which it is flowing in, minus the rate at which it is leaving.

The rate at which it accumulates is given by

$$\frac{\partial C}{\partial t} \text{ x volume of element} = \frac{\partial C}{\partial t} \Delta x \Delta y \Delta z$$

The rate at which it accumulates due to flow in the x direction is

$$\Delta y \Delta z \left(J_x - J_x + \left(\frac{\partial J_x}{\partial x} \right) \Delta x \right) = - \frac{\partial J_x}{\partial x} \Delta x \Delta y \Delta z$$

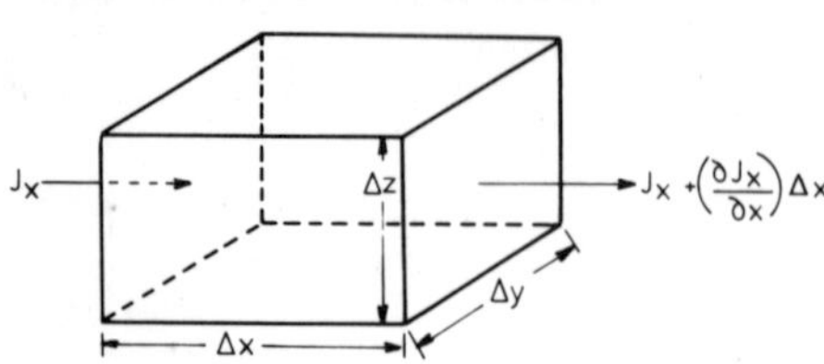

Fig. 2.1 Flow of diffusing material through a volume element

The rate at which it accumulates due to flow in all three dimensions is therefore

$$-\Delta x \Delta y \Delta z \left(\frac{\partial J_x}{\partial x} + \frac{\partial J_y}{\partial y} + \frac{\partial J_z}{\partial z} \right) = \frac{\partial C}{\partial t} \Delta x \Delta y \Delta z$$

Hence

$$\frac{\partial C}{\partial t} + \frac{\partial J_x}{\partial x} + \frac{\partial J_y}{\partial y} + \frac{\partial J_z}{\partial z} = 0 \tag{2.1}$$

In an isotropic medium, we can substitute from eqn. 1.1 and get

$$\frac{\partial C}{\partial t} = \frac{\partial}{\partial x}\left(D \frac{\partial C}{\partial x}\right) + \frac{\partial}{\partial y}\left(D \frac{\partial C}{\partial y}\right) + \frac{\partial}{\partial z}\left(D \frac{\partial C}{\partial z}\right) \tag{2.2}$$

If D is a constant, then this becomes

$$\frac{\partial C}{\partial t} = D\left(\frac{\partial^2 C}{\partial x^2} + \frac{\partial^2 C}{\partial y^2} + \frac{\partial^2 C}{\partial z^2}\right) \tag{2.3}$$

In many practical cases it is necessary to consider diffusion in only one dimension, i.e. there is a gradient only along the x axis. Then eqns. 2.2 and 2.3 reduce to

$$\frac{\partial C}{\partial t} = \frac{\partial}{\partial x}\left(D \frac{\partial C}{\partial x}\right) \tag{2.4}$$

and

$$\frac{\partial C}{\partial t} = D \frac{\partial^2 C}{\partial x^2} \tag{2.5}$$

Eqn. 2.5 is usually called Fick's second law, although it is often necessary to use eqn. 2.4 in the case of semiconductor diffusion. Fick's law can, of course, also be expressed in terms of cylindrical and spherical co-ordinates. For semi-conductor experiments, however, the cartesian form is invariably the most useful.

The mathematical solutions of the equation depend on the boundary conditions, which are determined, in turn, by the physical conditions of the experiment in question. In this chapter problems are considered in which the diffusion coefficient is a constant; cases for which D is a variable are considered later in the book, as they arise. The solutions given are not comprehensive, but cover most of the important situations that arise during semiconductor diffusion. More comprehensive accounts of the solution of the diffusion equation for almost any conceivable situation, are given in the books by Crank (1956)* and by Carslaw and Jaeger

* Crank, J., 'The mathematics of diffusion' (Oxford University Press, 1956)

(1959)†. When a problem can be solved in a fairly straight-forward manner, the solution is given. In the case of the more difficult problems, reference is made to where the solutions can be found. Three techniques are described:

(i) starting from the problem in which the solute is a very thin film, an extended source may be considered as being made up of many such films; the solution is then the sum of all the thin film solutions
(ii) Laplace transformation
(iii) method of separation of variables.

These are found to be the most useful techniques in this type of work.

Except for a section on anisotropy, it will be assumed that the diffusion flux and concentration gradient are both in the x direction, so that only the one-dimensional case for Fick's law need be considered. This means that any material diffusing in at the sides of the specimen (as opposed to the front and back faces) must be negligibly small. Two types of specimen are considered; the 'semi-infinite bar', in which the dimension in the direction of diffusion is much greater than the diffusion depth, and the 'thin' specimen, in which it is not. The semi-infinite approximation can usually be made in semiconductor work, and solutions for this case are therefore considered in somewhat more detail.

2.2 Semi-infinite specimens

2.2.1 Thin-film solution

Experimentally, this corresponds to the common case of plating a thin film of the diffusant on to the flat face of a semiconductor slice. The film then becomes the diffusion source. The problem is to solve Fick's law, subject to the condition that the total amount of plated substance is constant. It is simpler initially to consider the slightly different case of an infinite bar along the x direction with the thin film situated at the point ($x = 0$) perpendicular to the length of the bar. This problem is preferable because it is a symmetrical situation. The bar is raised to a high temperature for diffusion to take place. It is a simple matter to show that after a time, t, the distribution must be Gaussian:

$$C = \frac{A}{t^{1/2}} \exp\left(-\frac{x^2}{4Dt}\right) \tag{2.6}$$

The expression obeys eqn. 2.5 and the boundary conditions of the experiment; it is symmetrical with respect to $x = 0$, goes to zero as $x \rightarrow \pm\infty$ for $t > 0$, and for $t = 0$ vanishes everywhere, except for $x = 0$ where it is infinite. It remains to demonstrate

† Carslaw, H. S., Jaeger, J. C., 'Conduction of heat in solids' (Oxford University Press, 1959, 2nd edn.)

that the amount of diffusing material, M, say, is constant with time according to eqn. 2.6:

$$M = \int_{-\infty}^{\infty} \frac{A}{t^{1/2}} \exp \frac{-x^2}{4Dt} \, dx \tag{2.7}$$

changing the variable to $u^2 = x^2/4Dt$, eqn. 2.7 becomes

$$M = \int_{-\infty}^{\infty} 2AD^{1/2} \exp(-u^2) \, du \tag{2.8}$$

and using the fact that $\int_{-\infty}^{\infty} \exp(-u^2) \, du = \pi^{1/2}$, we have $M = 2A(\pi D)^{1/2}$. Thus the amount of solute in the bar is constant with respect to time, and the first condition is obeyed. The complete thin-film solution is therefore

$$C = \frac{M}{2(\pi Dt)^{1/2}} \cdot \exp\left(\frac{-x^2}{4Dt}\right) \tag{2.9}$$

Fig. 2.2 shows eqn. 2.9 for various times.

Let us now return to the more realistic case in semiconductor diffusion, of a thin film plated onto a semi-infinite slice. Since the 'infinite' problem is symmetrical, we can slice it down the middle and merely take the solution for positive x. Physically this is reasonable since Fig. 2.2 shows that at $x = 0$, dC/dx is zero at all times. From Fick's first law this means that there is no flow of atoms across the $x = 0$ boundary,

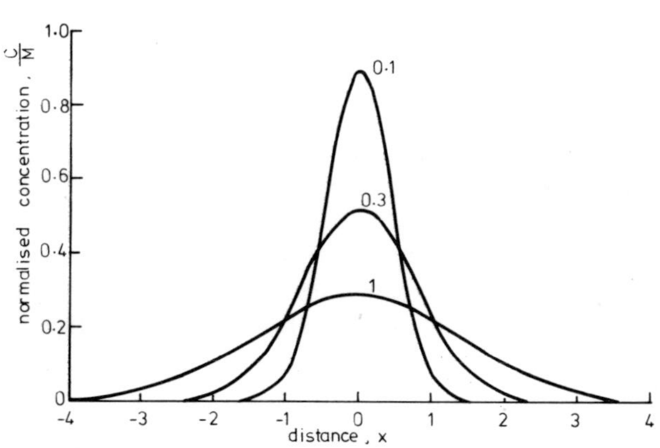

Fig. 2.2 Thin-film solutions to the diffusion equation for different diffusion times, plotted as C/M against x. The numbers on the curves refer to different values of the quantity Dt

and the two halves of the problem are therefore effectively insulated from each other. We must allow for the fact that in the infinite problem, half of the mass goes in one direction and half in the other. If exactly the same result is to be obtained in the region $x > 0$ in the two cases, this means that the amount of diffusant plated onto the semi-infinite specimen must be only half that used in the infinite bar. Let the amount of material plated on to the semi-infinite specimen be α, then

$$C = \frac{\alpha}{(\pi Dt)^{1/2}} \exp\left(\frac{-x^2}{4Dt}\right) \tag{2.10}$$

We are now in a position to be more precise about what is meant by 'thin' and 'semi-infinite'. A semi-infinite specimen is one for which the dimension in the diffusion direction is greater than the diffusion depth. Since profiles of the type 2.7 in principle extend to infinity, it is necessary to have some practical definition of penetration depth. If solute reaches the end of the bar, it might immediately evaporate from it, or it might be reflected back into the bar, depending on the conditions at the end. In either case, the profile must be disturbed from that of eqn. 2.7. Provided the concentration of solute at the far end is small, this will be a negligible effect, and the bar can reasonably be called semi-infinite. Define this condition by stipulating that the profile must fall to 1 percent of the value it has at $x = 0$, i.e.

$$\exp\left(\frac{-l^2}{4Dt}\right) = 0{\cdot}01$$

where l is the minimum length for a semi-infinite specimen. This gives a value for l of $4{\cdot}3 \times (Dt)^{1/2}$ and underlines the importance of the length $(Dt)^{1/2}$, which is the natural unit of length in this subject. For many diffusion experiments the diffusion time is selected to make the penetration depth small compared to the thickness of the semiconductor slice (which might be 0·5 mm), so the semi-infinite approximation is very often valid.

2.2.2 Pair of semi-infinite specimens

Consider a pair of semi-infinite solid bars, one occupying the space $-\infty < x < 0$, the other occupying $0 < x < \infty$. They are joined at the plane $x = 0$, which is perpendicular to the length of the bars. The first bar is homogeneously doped with some solute to a concentration C', the other contains no solute. Apart from this the two bars are identical, as in Fig. 2.3. The boundary conditions may therefore be stated:

$$\left.\begin{array}{ll} C = C' & \text{for } x < 0 \\ C = 0 & \text{for } x < 0 \end{array}\right\} \quad t = 0$$

The simplest way to solve the problem is to imagine the left hand distribution to be made up of an infinite number of thin films. We can then sum the contributions

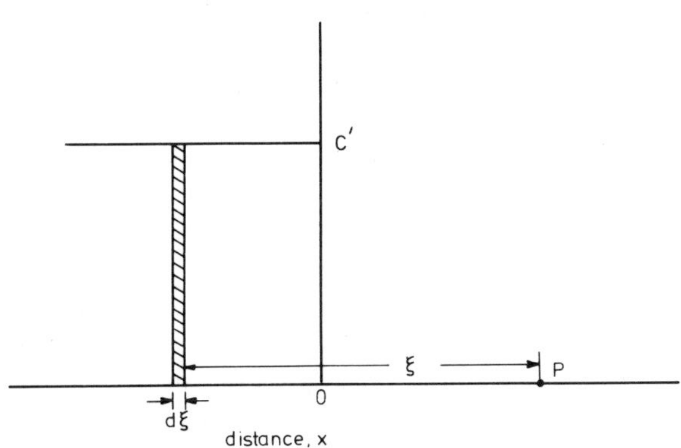

Fig. 2.3 Initial solute concentration in a pair of semi-infinite bars

from the films using eqn. 2.9. Consider a general point, P, at x in Fig. 2.3, and one of these films, a distance ξ from P. The thickness of the film is $d\xi$, and it contains an amount of solute $C'd\xi$. From eqn. 2.9 the amount of solute at P after time t is given by

$$\frac{C'd\xi}{2(\pi Dt)^{1/2}} \exp\left(-\frac{\xi^2}{4Dt}\right)$$

The amount of solute at P due to all the films is the sum from $\xi = x$ to $\xi = +\infty$, which includes all films, i.e.

$$C = \int_x^\infty \frac{C'}{2(\pi Dt)^{1/2}} \exp\left(\frac{-\xi^2}{4Dt}\right) d\xi \tag{2.11}$$

Changing the variable to $\eta = \xi/2(Dt)^{1/2}$ this becomes

$$C = \frac{C'}{\pi^{1/2}} \int_{x/2(Dt)^{1/2}}^\infty \exp(-\eta^2)\, d\eta \tag{2.12}$$

We now have the distribution in the form of a standard function. The error function is defined as

$$\operatorname{erf} z = \frac{2}{\pi^{1/2}} \int_0^z \exp(-\eta^2)\, d\eta \tag{2.13}$$

Clearly this function has the properties erf$(z) = -$erf$(-z)$ and erf $(0) = 0$. It can also be shown that erf $(\infty) = 1$. Now eqn. 2.12 can be written

$$C = \frac{C'}{2} \cdot \frac{2}{\pi^{1/2}} \int_z^\infty \exp(-\eta^2)\, d\eta$$

where $z = x/2(Dt)^{1/2}$, and

$$\frac{2}{\pi^{1/2}} \int_z^\infty \exp(-\eta^2)\, d\eta = \frac{2}{\pi^{1/2}} \int_0^\infty \exp(-\eta^2)\, d\eta - \frac{2}{\pi^{1/2}} \int_0^z \exp(-\eta^2)\, d\eta$$

$$= 1 - \text{erf}\, z = \text{erfc}\,(z)$$

This last function is called the error function complement. Eqn. 2.12 can therefore be written

$$C = \frac{C'}{2} \text{erfc} \frac{x}{2(Dt)^{1/2}} \tag{2.14}$$

Both the erf and erfc functions have been extensively calculated; a range of values is shown in Table 2.1. Fig. 2.4 is a plot of eqn. 2.14 using $2(Dt)^{1/2}$ as the unit of length. The concentration drops to about 1 percent of C' at about $x = 3{\cdot}6 \times (Dt)^{1/2}$.

Table 2.1 Values of the error function and error function complement

(Comprehensive tables of the function can be found in 'Tables of the probability function' Works Project Association, New York, 1941)

x	erf *x*	erfc *x*	*x*	erf *x*	erfc *x*
0	0	1·0000	0·70	0·6778	0·3222
0·05	0·0564	0·9436	0·75	0·7112	0·2888
1·10	0·1125	0·8875	0·80	0·7421	0·2579
0·15	0·1680	0·8320	0·90	0·7969	0·2031
0·20	0·2227	0·7773	1·0	0·8427	0·1573
0·25	0·2763	0·7237	1·1	0·8802	0·1198
0·30	0·3286	0·6714	1·2	0·9103	0·0897
0·35	0·3794	0·6206	1·4	0·9523	0·0477
0·40	0·4284	0·5716	1·6	0·9763	0·0237
0·45	0·4755	0·5245	1·8	0·9891	0·0109
0·50	0·5205	0·4795	2·0	0·9953	0·0047
0·55	0·5633	0·4367	2·4	0·9993	0·0007
0·60	0·6039	0·3961	2·8	0·9999	0·0001
0·65	0·6420	0·3580			

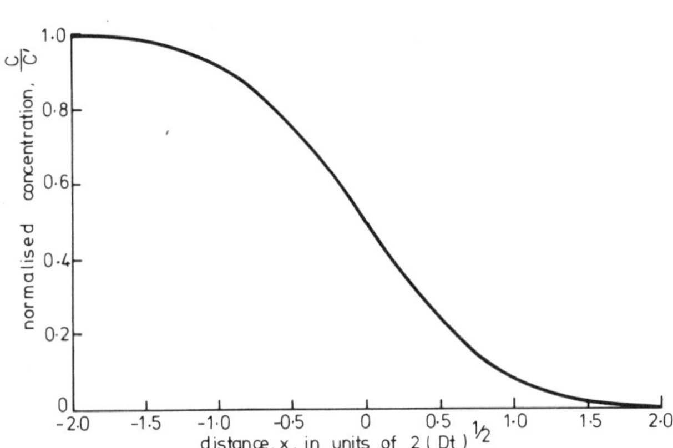

Fig. 2.4 Error function complement curve plotted using $2(Dt)^{1/2}$ as the unit of length

Suppose both bars are doped with the diffusant so that at time $t = 0$ the doping for $x < 0$ is C_1', and that for $x > 0$ is C_2'. Let C_1' be greater than C_2'. There are two ways to look at this problem, and it is instructive to consider both. We can take the view that we have two problems of the type presented by Fig. 2.3, with one diffusing to the left and one diffusing to the right. The solution must therefore be the sum of the two simple solutions:

$$C = \frac{C_1'}{2}\left(1 - \operatorname{erf}\frac{x}{2(Dt)^{1/2}}\right) + \frac{C_2'}{2}\left(1 + \operatorname{erf}\frac{x}{2(Dt)^{1/2}}\right) \tag{2.15}$$

where the positive sign in the second bracket indicates right-to-left diffusion.

On the other hand, we could solve the problem by moving the zero of concentration from $C = 0$ to $C = C_2'$. The problem then simplifies to Fig. 2.3 again. The solution is now essentially eqn. 2.14

$$C - C_2' = \frac{C_1' - C_2'}{2} \operatorname{erfc}\frac{x}{2(Dt)^{1/2}} \tag{2.16}$$

It is a simple matter to show that eqns. 2.15 and 2.16 are the same solution.

An initial distribution such as that shown in Fig. 2.3 might be approximately realised by growing a thick doped crystal on to an undoped thick substrate made of the same material. The growth might take place either from a molten phase or from a vapour phase. It is important to realise, however, that the step function in concentration cannot be achieved exactly because crystal growth processes take place at elevated temperature, so that some diffusion must take place during layer growth. Diffusion during growth of a layer is, in fact, considered as a separate case in Section 2.2.6.

2.2.3 Constant surface concentration

One of the features of the erfc curve is that the value of the function at $x = 0$ is unity for all time. Reference to eqn. 2.14 shows that at $x = 0$, the concentration is always $\frac{1}{2}C'$. It follows that for positive values of x, the problem described in Section 2.2.2 is identical to one in which the surface concentration is kept at $\frac{1}{2}C'$ for all time, by some other means. This set of conditions is common for diffusion in semiconductors. A slice of a semiconductor is kept in a container at high temperature, surrounded by a vapour of the material to be diffused. At the temperature and vapour pressure of the experiment, the solute has a solubility in the semiconductor of C_0, say. Let us assume that the surface of the semiconductor achieves this concentration of solute immediately. Then providing the vapour pressure does not change throughout the experiment, the surface concentration will be maintained at C_0 and the diffusion profile will be of the form

$$C = C_0 \operatorname{erfc} \frac{x}{2(Dt)^{1/2}} \tag{2.17}$$

where C_0 corresponds to $\frac{1}{2}C'$ in eqn. 2.14 and the semiconductor occupies the space $0 < x < \infty$.

Essentially, the same solution is correct for one or two similar situations. Suppose the sample is homogeneously doped to a level of C_1 and an external vapour is used to change the surface concentration to a different concentration, C_2. Then by the argument leading to eqn. 2.16, the profile is

$$C - C_1 = (C_2 - C_1) \operatorname{erfc} \frac{x}{2(Dt)^{1/2}} \tag{2.18}$$

Eqn. 2.18 can also be used to cover the case of out-diffusion when a sample which is initially uniformly doped to a level C_1, say, is heated in a vacuum. In this case, providing the dopant which is diffusing out is pumped away, so that no solute vapour collects near the specimen, we have $C_2 = 0$, and the solution is

$$C - C_1 = -C_1 \operatorname{erfc} \frac{x}{2(Dt)^{1/2}}$$

which becomes

$$C = C_1 \operatorname{erf} \frac{x}{2(Dt)^{1/2}} \tag{2.19}$$

The constant surface concentration case, which is probably the most important from the point of view of semiconductor work, also provides an instructive example of a Fick's law problem that can be simply solved using Laplace transforms. The Laplace transform of a function $f(t)$ is defined as $\bar{f}(p)$, where

$$\bar{f}(p) = \int_0^\infty f(t) \exp(-pt)\, dt \tag{2.20}$$

and p is some number large enough to make the integral finite. Some examples of Laplace transforms are shown in Table 2.2.

To restate the problem, we require a solution to

$$\frac{\partial C}{\partial t} = D\frac{\partial^2 C}{\partial x^2} \tag{2.21}$$

subject to the conditions

$$C = 0 \quad x > 0 \quad \text{for} \quad t = 0 \tag{2.22}$$

$$C = C_0 \quad x = 0 \quad \text{for} \quad t > 0 \tag{2.23}$$

Multiplying both sides of eqn. 2.21 by $\exp(-pt)$ and integrating from $t = 0$ to $t = \infty$ gives

$$\int_0^\infty \frac{\partial C}{\partial t} \exp(-pt)\, dt = D \int_0^\infty \frac{\partial^2 C}{\partial x^2} \exp(-pt)\, dt \tag{2.24}$$

Taking the left-hand side, and integrating by parts

$$\int_0^\infty \frac{\partial C}{\partial t} \exp(-pt)\, dt = [C \exp(-pt)] + p \int_0^\infty C \exp(-pt)\, dt = p\bar{C}$$

Table 2.2 Table of Laplace transforms

$f(t)$	$\bar{f}(p)$
1	$\dfrac{1}{p}$
$\exp(at)$	$\dfrac{1}{p-a}$
$\sin \omega t$	$\dfrac{\omega}{p^2+\omega^2}$
$\cos \omega t$	$\dfrac{p}{p^2+\omega^2}$
$\exp\left\{-\left(\dfrac{p}{D}\right)^{1/2} x\right\}$	$\dfrac{x}{2(\pi D t^3)^{1/2}} \exp\left(-\dfrac{x^2}{4Dt}\right)$
$\dfrac{1}{p} \exp\left\{-\left(\dfrac{p}{D}\right)^{1/2} x\right\}$	$\operatorname{erfc} \dfrac{x}{2(Dt)^{1/2}}$

where $\bar{C}$ is the Laplace transform of C. The term in the square brackets is zero at $t = 0$ from eqn. 2.22 and also at $t = \infty$. Taking the right-hand side of eqn. 2.24:

$$D \int_0^\infty \frac{\partial^2 C}{\partial x^2} \exp(-pt)\, dt = D \frac{\partial^2}{\partial x^2} \int_0^\infty C \exp(-pt)\, dt = D \frac{\partial^2 \bar{C}}{\partial x^2}$$

where the small liberty of reversing the order of integration and differentiation has been taken. Eqn. 2.24 becomes

$$p\bar{C} = D \frac{d^2 \bar{C}}{dx^2} \tag{2.25}$$

The partial differentials are now full. Applying the same treatment to the boundary conditions, eqn. 2.23 becomes

$$\bar{C} = \int_0^\infty C \exp(-pt)\, dt = \int_0^\infty C_0 \exp(-pt)\, dt \qquad \text{at } x = 0$$

$$\text{i.e.} \quad \bar{C} = \frac{C_0}{p} \quad \text{at } x = 0 \tag{2.26}$$

The solution of eqn. 2.25 subject to condition 2.26 is straightforward, and becomes

$$\bar{C} = \frac{C_0}{p} \exp\left(-\left(\frac{p}{D}\right)^{1/2} x\right) \tag{2.27}$$

Comparison of eqn. 2.27 with the (carefully selected) transforms given in the Table 2.2 show it to be the transform of

$$C = C_0 \operatorname{erfc} \frac{x}{2(Dt)^{1/2}} \tag{2.28}$$

This problem comes out particularly easily because it happens to give a standard transform straight away. More skill is needed in more difficult cases.

It is of some interest to consider the rate at which solute enters the solid from the vapour phase. Let J_0 be the rate of flow of solute across the vapour – solid interface at $x = 0$, then

$$J_0 = \left(D \frac{\partial C}{\partial x}\right)_{x=0} = \left[-D \frac{\partial}{\partial x} C_0 \left(1 - \operatorname{erf} \frac{x}{2(Dt)^{1/2}}\right)\right]_{x=0}$$

$$= \left[D \frac{\partial}{\partial x} C_0 \operatorname{erf} \frac{x}{2(Dt)^{1/2}}\right]_{x=0} = \left[\frac{D_0 C_0}{(\pi D t)^{1/2}} \cdot \exp\left(\frac{-x^2}{(4Dt)}\right)\right]_{x=0}$$

$$= \frac{DC_0}{(\pi D t)^{1/2}} \tag{2.29}$$

The amount of material that has entered in time t is given by integrating J_0, and comes to

$$M = 2C_0 \left(\frac{Dt}{\pi}\right)^{1/2} \tag{2.30}$$

Eqn. 2.30 is particularly relevant to the case of diffusion in a closed vessel from the vapour phase. For the boundary condition of constant surface concentration to be maintained, it is also necessary for the ambient vapour pressure of the diffusing material to be maintained, since it is this pressure which determines the surface concentration. The vapour pressure must be decreasing, because the vapour is continually losing material to the solid, according to eqn. 2.30. It follows that unless the material lost to the solid is negligibly small compared to the total amount in the vapour phase, eqn. 2.17 will not be the solution.

2.2.4 Diffusion from a thick film

This is diffusion into a semiconductor, from a film deposited on the surface, when the thickness of the film cannot be ignored. This amounts to saying that the film thickness is not small compared to $(Dt)^{1/2}$. Again it is simpler initially to consider the case of the film sandwiched between two semi-infinite rods. Let the film occupy the space $-h < x < h$ and the rods occupy $-\infty < x < -h$ and $h < x < \infty$ respectively so that the problem is symmetrical about the $x = 0$ plane. We can follow exactly the same procedure as in Section 2.2.2. The thick film can be considered as being made up of many thin films, each of which contribute a concentration at a general point x. These contributions must in this case be summed between $(x + h)$ and $(x - h)$, giving

$$C = \int_{(x-h)}^{(x+h)} \frac{C'}{2(\pi Dt)^{1/2}} \exp\left(-\frac{\xi^2}{4Dt}\right) d\xi \tag{2.31}$$

(compare with eqn. 2.11). Again making the substitution $\eta = \xi/2(Dt)^{1/2}$, eqn. 2.31 gives

$$C = \frac{C'}{\pi^{1/2}} \int_{x-h/2(Dt)^{1/2}}^{x+h/2(Dt)^{1/2}} \exp(\eta^2)\, d\eta \tag{2.32}$$

Splitting up the integral and making use of the property $\text{erf}\,(z) = -\text{erf}\,(-z)$, eqn. 2.32 becomes

$$C = \frac{C'}{\pi^{1/2}} \int_0^{h+x/2(Dt)^{1/2}} \exp(-\eta^2)\, d\eta - \frac{C'}{\pi^{1/2}} \int_0^{x-h/2(Dt)^{1/2}} \exp(-\eta^2)\, d\eta \tag{2.33}$$

$$C = \frac{C'}{2}\left(\text{erf}\,\frac{h+x}{2(Dt)^{1/2}} + \text{erf}\,\frac{h-x}{2(Dt)^{1/2}}\right) \tag{2.34}$$

Once again the solution is symmetrical about $x = 0$, with $(\partial C/\partial x)$ therefore zero at $x = 0$. There can be no flow across the origin, and the problem can be split down the middle. The solution 2.34 for positive values of x therefore describes the experiment in which a layer of thickness, h, is diffused into a semi-infinite specimen. The total amount of solute must always be equal to $M = hC'$. This is confirmed by integrating eqn. 2.34 in the range $0 < x < \infty$. The profile is shown in Fig. 2.5. The surface concentration at $x = 0$ decreases as

$$C(0, t) = C' \operatorname{erf} \frac{h}{2(Dt)^{1/2}}$$

2.2.5 Diffusion into an evaporating specimen

When diffusion from the vapour phase takes place at a temperature close to the melting-point of the semiconductor crystal, evaporation from the crystal can occur on quite a large scale. Any crystal will require a vapour pressure of its own elements, to be in equilibrium at a high temperature. If the diffusion takes place in a closed vessel, the crystal will evaporate until the correct vapour pressure has been built up. In open systems, the evaporating material will tend to be carried away by the flow of the vapour, and the required vapour pressure may not be achieved. In this case the crystal will continue to evaporate. The same effect; removing of surface material, sometimes occurs by chemical means. It is not uncommon for the vapour to react with a semiconductor in such a way as to etch it.

Both of these examples resolve into a problem of diffusion of some solute from

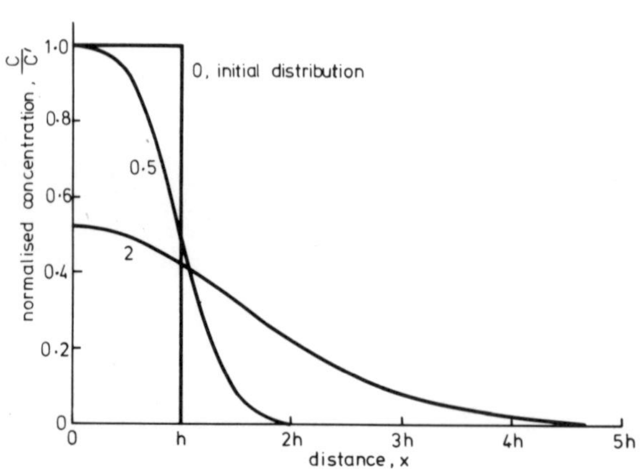

Fig. 2.5 Thick-film solutions to the diffusion equation for different diffusion times. The numbers on the curves refer to different values of parameter $(2/h)(Dt)^{1/2}$

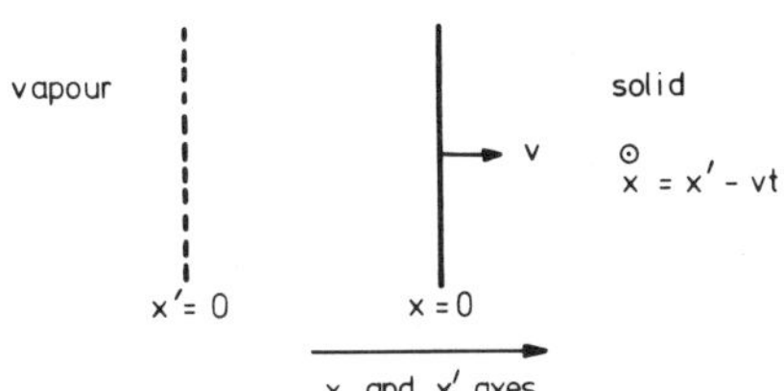

Fig. 2.6 Diffusion into an evaporating specimen

the vapour phase into a solid, when the interface is moving with velocity v. We shall assume instantaneous surface equilibrium so that at all times the concentration at the moving interface is the equilibrium concentration, C_0. The situation is shown diagrammatically in Fig. 2.6. It may be seen that a point within the solid might be described by either of two sets of axes, one with a static zero at the interface position at time $t = 0$, call this the x' axis, and the other taking the moving interface as zero. Call this the x axis. They are related by $x = x' - vt$.

Initially stating the problem with reference to the static axis, it is necessary to solve the diffusion equation subject to:

$$C(x', 0) = 0 \quad \text{for} \quad x' > 0 \quad \text{at} \quad t = 0$$

$$C(x', t) = C_0 \quad \text{for} \quad x' = vt \text{ at all time } t.$$

If the variable is changed from x' to x, i.e. if we change to the moving co-ordinates, Fick's law becomes

$$D \frac{\partial^2 C}{\partial x^2} = \frac{\partial C}{\partial t} - v \frac{\partial C}{\partial x} \tag{2.35}$$

The problem has been solved by Kucher (1961)*. The profile has the form

$$C = \frac{C_0}{2} \left(\operatorname{erfc} \frac{x + vt}{2(Dt)^{1/2}} + \exp - \frac{vx}{D} \operatorname{erfc} \frac{x - vt}{2(Dt)^{1/2}} \right) \tag{2.36}$$

If the velocity is zero, the two frames of reference are identical, i.e. $x = x'$ and eqn. 2.36 becomes

$$C = C_0 \operatorname{erfc} \frac{x}{2(Dt)^{1/2}},$$

as expected.

* Kucher, T. I., *Sov. Phys. – Solid State*, 1961, **3**, p. 401

An interesting situation occurs in the case of high evaporation rate and low diffusion coefficient; we can then write $vt \gg 2(Dt)^{1/2}$. Now since most of any diffusion profile will be within a few times the length $(Dt)^{1/2}$, we can say that we are only interested in values of x of less than $4(Dt)^{1/2}$, i.e. $x \ll vt$. Substituting these two ideas into eqn. 2.36, the first erfc function tends to zero and the second goes to 2. Hence an approximate solution is

$$C = C_0 \exp - \frac{vx}{D} \tag{2.37}$$

which gives the surprising result that the profile is independent of time. In physical terms this is because the movement of a diffusing impurity front slows down with increasing penetration, and there will be a certain depth at which the movement is equal to the velocity of the interface. At this stage a steady state is set up, and the depth of penetration cannot be increased. The restriction $vt \gg 2(Dt)^{1/2}$ is quite strong, although for a finite velocity it must be satisfied after a sufficiently long time. In fact the steady-state profile, eqn. 2.37, could have been obtained directly by simply putting $\partial C/\partial t = 0$ in eqn. 2.35.

The effect was investigated by Batdorf and Smits (1959)*, diffusing impurities of groups III and V into silicon, at temperatures in the range 1200°C–1350°C. They carried out their diffusions in an apparatus that was continually pumped, so that silicon was evaporating all the time. The vapour pressure of the diffusing species was maintained by having a source within the apparatus that was also continually evaporating. The temperature of the source was kept sufficiently high to achieve the required vapour pressure. Their evaporation rates from the silicon sample were as high at 150 $\mathrm{nms^{-1}}$ for the higher temperatures, and they claimed that their diffusions followed the form of eqn. 2.37.

Consider an evaporation rate of 100 $\mathrm{nms^{-1}}$ in an experiment lasting for 10^5 s. The restriction $vt \gg 2(Dt)^{1/2}$ would then require a diffusion coefficient given by $D \ll \frac{1}{4}v^2 t$, i.e. $D \ll 2{\cdot}5 \times 10^{-6}\ \mathrm{m^2 s^{-1}}$, and eqn. 2.37 would be correct for distances in the crystal much less than 0·1 mm. The effect may well be of significance in diffusion into compound semiconductors which can have very large vapour pressures at high temperatures. Close to its melting point, GaP, for instance, has an equilibrium vapour pressure of phosphorus in the region of $5 \times 10^6\ \mathrm{Nm^{-2}}$ (50 atmospheres!). It is certainly a common experience to find that a good deal of evaporation has taken place when one is diffusing into III–V semiconductors. It is quite possible that some evaporation of the surface is desirable from the point of view of the semiconductor device, because of the clean surface revealed.

2.2.6 Diffusion from a growing epitaxial layer

For modern semiconductor devices, the material is often required in the form of a thin single-crystal layer, on a substrate either of the same or of some other material.

* Batdorf, R. L., and Smits, F. M.: *J. Appl. Phys.*, 1959, **30**, p. 259

Layers are usually deposited from the vapour phase, and the substrates are single crystals. If required, a deposited layer may have impurities incorporated by mixing in some suitable vapour. The substrate is also likely to contain dopant atoms, and the deposited layer may be of the same or of opposite impurity type to that of the layer. This process is called epitaxial growth. Epitaxial deposition temperatures are often quite close to the melting-point of the semiconductor being grown, and the diffusivities of the various impurities may be quite large. The question therefore arises of impurities diffusing out of the depositing layer into the substrate and vice versa. This problem has been considered in some detail by Rice (1964)*. He assumed the substrate could be treated as a semi-infinite sample.

The situation is essentially the same as that depicted in Fig. 2.6 except that the interface is now moving into the vapour, rather than into the solid. Consider initially the case of the substrate being undoped and the layer being put down with doping concentration C_0. As with the previous example, we choose the constant concentration layer as the origin of co-ordinates, so that $x = 0$ defines the growing surface and the layer-substrate interface recedes from the origin with velocity, v. The problem then resolves into one which differs from that of Section 2.2.5 only in the sign of v, i.e. we have growth instead of evaporation. The diffusion equation becomes

$$D\frac{\partial^2 C}{\partial x^2} = \frac{\partial C}{\partial t} + v\frac{\partial C}{\partial x} \tag{2.38}$$

subject to the conditions:

$C = C_0$ at $x = 0$ for all t

$C = 0$ as $x \to \infty$

$C = 0$ for $x > 0$ at $t = 0$.

The only difference between the solution to this problem and eqn. 2.36 is that the sign of v is changed, i.e.

$$C = \frac{C_0}{2}\left(\operatorname{erfc}\frac{x - vt}{2(Dt)^{1/2}} + \exp\frac{vx}{D}\operatorname{erfc}\frac{x + vt}{2(Dt)^{1/2}}\right) \tag{2.39}$$

The second case to consider is that of an undoped layer being put down on an homogeneously doped semi-infinite substrate. The problem is a little more difficult, because we need to allow for the fact that some of the dopant that diffuses into the layer from the substrate will diffuse straight through, and evaporate from the growth surface. There will be a net loss of dopant due to evaporation if the rate at which dopant atoms leave the surface at $x = 0$ is greater than the rate at which they

* Rice, W. R., *Proc. IEEE*, 1964, **52**, p. 284

join it from the ambient atmosphere. Now most epitaxial-growth systems have a steady flow of reactants through the equipment, so that any vapour leaving the specimen and entering the vapour is removed. The vapour pressure of dopant is therefore approximately zero, and there is a net loss. The simplest assumption to make, under these circumstances, is that the loss at the growing surface is proportional to the concentration at the surface, with a constant of proportionality K. Mathematically, this means solving eqn. 2.38 with the conditions:

$C = C_0$ for all x at $t = 0$.

$C = C_0$ as $x \to \infty$ for all time

An additional rate equation must be satisfied at the surface:

$$D \cdot \frac{\partial C}{\partial x} = (K + v)C(0, t) \quad \text{at } x = 0 \tag{2.40}$$

This condition is not immediately obvious. The effect of choosing the growing surface as the origin is to give all the diffusing atoms a constant velocity v in the x-direction. Consider unit area drawn in the growing surface. For continuity of flux, the flow of atoms on the solid side must equal that on the vapour side of the unit area. The former is due to the diffusion flux, minus the flux back into the solid due to the superimposed velocity i.e.

$$J = D\left(\frac{\partial C}{\partial x}\right)_{x=0} - vC(0, t),$$

where, according to our chosen set of co-ordinates, J is a flux in the negative x-direction. The latter flow of atoms, on the vapour side, is due to evaporation, $KC(0, t)$. Equating these flows gives eqn. 2.40.

The solution is rather complicated:

$$C = C_0 - \frac{C_0}{2}\left(\operatorname{erfc}\frac{x - vt}{2(Dt)^{1/2}} + \frac{K + v}{K}\exp\frac{vx}{D}\operatorname{erfc}\frac{x + vt}{2(Dt)^{1/2}}\right)$$

$$+ C_0\left(\frac{2K + v}{2K}\right)\exp\frac{(K + v)(x + Kt)}{D}\operatorname{erfc}\frac{x + (2K + v)t}{2(Dt)^{1/2}} \tag{2.41}$$

As $K \to 0$, eqn. 2.41 reduces to the simple solution

$$C = \frac{C_0}{2}\left(1 + \operatorname{erf}\frac{x - vt}{2(Dt)^{1/2}}\right) \tag{2.42}$$

Physically this means that if there is no loss from the growing surface by evaporation, then the diffusing species does not know that there is a surface there. The epitaxial layer behaves in this case like a semi-infinite solid. The model becomes a little unrealistic at this point, however, since in the absence of evaporation from the growing interface, reflection of the profile would occur.

The situation must be generalised further in order to get closer to a real laboratory crystal growth. Usually there will be impurities both in the layer and in the substrate, concentrations C_1 and C_2, with diffusion constants D_1 and D_2. The solution for this case is a combination of eqns. 2.39 and 2.41. For further details the the reader is referred to Rice's paper.

2.2.7 Diffusion through a layer of one material into a semi-infinite specimen of another material

This is an important case in semiconductor technology. It is common practice to mask silicon with a layer of silicon oxide to prevent diffusion taking place in regions of the semiconductor where it is not required. Holes are made in the oxide layer to allow diffusion to occur in the right places. The diffusing material does, of course, enter the oxide, and it is therefore necessary to consider how thick the oxide layer must be to prevent an appreciable amount of dopant entering the semiconductor from the oxide. In other experiments the oxide layer is used in order to obtain some special doping profile, and the dopant is required to enter the semiconductor. Diffusion through a layer has been studied both theoretically and experimentally by Sah *et al.* *, diffusing phosphorus into silicon.

Let the diffusion coefficient in the layer be D_1 and that in the semiconductor be D_2. Take the layer thickness to be a and the semiconductor surface to be the zero of co-ordinates, so that the oxide-ambient interface is at $x = -a$. Then

$$\left.\begin{aligned} D_1 \frac{\partial^2 C_1}{\partial x^2} &= \frac{\partial C_1}{\partial t} \quad \text{for } -a < x < 0 \\ D_2 \frac{\partial^2 C_2}{\partial x^2} &= \frac{\partial C_2}{\partial t} \quad \text{for } x > 0 \end{aligned}\right\} \tag{2.43}$$

There is also the condition that the flux of atoms leaving the layer at $x = 0$ is equal to that joining the semiconductor, i.e.

$$D_1 \frac{\partial C_1}{\partial x} = D_2 \frac{\partial C_2}{\partial x} \quad \text{at } x = 0 \tag{2.44}$$

If the diffusion is from the vapour phase, the surface at $x = -a$ will be maintained at some constant value;

$$C_1 = C_0 \text{ at } x = -a \text{ for all time} \tag{2.45}$$

If it is assumed that there is no discontinuity in concentration at the boundary between the two solid mediums then

$$C_1 = C_2 \text{ at } x = 0 \text{ for all time} \tag{2.46}$$

and also

$$C_2 \to 0 \text{ as } x \to \infty \tag{2.47}$$

* Sah, C. T., Sello, H., and Tremere, D. A., *J. Phys. & Chem. Solids*, 1959, **11**, p. 228

Eqns. 2.43–2.47 are solved in Carslaw and Jaeger using the method of Laplace transforms. The solutions are

$$C_1 = C_0 \sum_{j=0}^{\infty} \left(\frac{1-\mu}{1+\mu}\right)^j \left\{\operatorname{erfc} \frac{a(2j+1)+x}{2(D_1 t)^{1/2}} - \frac{1-\mu}{1+\mu} \operatorname{erfc} \frac{a(2j+1)-x}{2(D_2 t)^{1/2}}\right\} \tag{2.48}$$

$$C_2 = \frac{2\mu C_0}{1+\mu} \sum_{j=0}^{\infty} \left(\frac{1-\mu}{1+\mu}\right)^j \operatorname{erfc}\left\{\frac{a(2j+1)}{2(D_1 t)^{1/2}} + \frac{x}{2(D_2 t)^{1/2}}\right\} \tag{2.49}$$

where

$$\mu^2 = \frac{D_1}{D_2}$$

The equations are greatly simplified for the case of $a^2 > 4D_1 t$. The first term in the summation, for $j = 0$, is then a good approximation because erfc x decreases very rapidly for increasing x. Eqns. 2.48 and 2.49 then become

$$C_1 \simeq C_0 \left(\operatorname{erfc} \frac{a+x}{2(D_1 t)^{1/2}} - \frac{1-\mu}{1+\mu} \operatorname{erfc} \frac{a-x}{2(D_1 t)^{1/2}}\right) \tag{2.50}$$

$$C_2 \simeq \frac{2\mu C_0}{1+\mu} \operatorname{erfc}\left(\frac{a}{2(D_1 t)^{1/2}} + \frac{x}{2(D_2 t)^{1/2}}\right) \tag{2.51}$$

Experiments diffusing phosphorus into silicon through a silicon oxide film have given a value of 0·34 for μ, i.e. diffusion in the oxide is slow compared to that in the semiconductor. A graph of eqns. 2.50 and 2.51 is shown in Fig. 2.7a with $D_1 < D_2$.

In general the concentrations in the two will not be equal at their interface and $C_2(0, t) = \mathrm{k}C_1(0, t)$, where k is the segregation coefficient. An example of the solution for $\mathrm{k} < 1$ is shown in Fig. 2.7b. In this case, again with the restriction that $a^2 > 4D_1 t$, the solutions are

$$C_1 \simeq C_0 \left(\operatorname{erfc} \frac{a+x}{2(D_1 t)^{1/2}} - \frac{\mathrm{k}-\mu}{\mathrm{k}+\mu} \cdot \operatorname{erfc} \frac{a-x}{2(D_1 t)^{1/2}}\right) \tag{2.52}$$

$$C_2 \simeq \frac{2\mathrm{k}\mu}{\mathrm{k}+\mu} C_0 \operatorname{erfc}\left(\frac{a}{2(D_1 t)^{1/2}} + \frac{x}{2(D_2 t)^{1/2}}\right) \tag{2.53}$$

Another feature of this type of diffusion is that the surface concentration of the semiconductor is very much less than it would have been without the protective film. This is quite clear in Fig. 2.7 and is always true if $\mathrm{k} \leqslant 1$. In some semiconductors it is difficult to achieve very low surface concentrations, and diffusion through an oxide film has been successfully used to do this. The oxide film can be dissolved off the semiconductor at the end of the experiment.

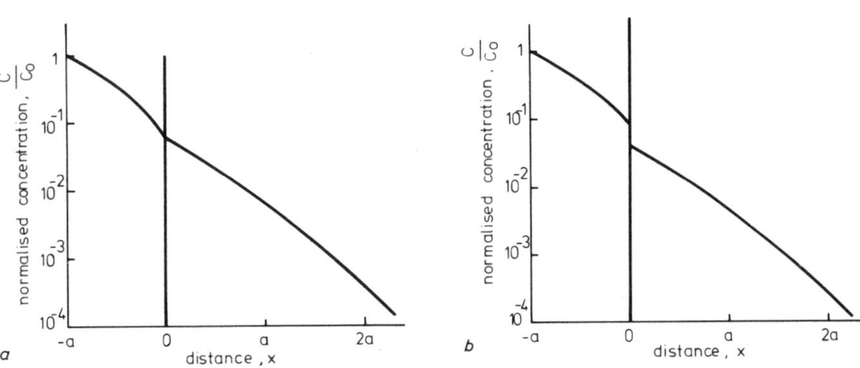

Fig. 2.7 Diffusion through a layer of material, thickness a, into a semi-infinite specimen of another material. Curves plotted for $a^2 = 6D_1 t$ and $\mu = \frac{1}{2}$

a Segregation coefficient, $k = 1$
b Segregation coefficient, $k < 1$

2.2.8 Diffusion in a potential gradient

If a field is applied to a sample in which atoms are diffusing, then Fick's law must be modified to allow for this. The field might be either mechanical or electrical. Diffusion under the action of gravity, for instance, is an example of the former. (One might say, of course, that all diffusion is under the action of gravity. It is simply a matter of whether the movement of atoms due to gravity is comparable to that due to the concentration gradient. In most cases of diffusion in solids it is negligible, but it is possible to enhance the effect by carrying out the work in a centrifuge, thereby helping gravity a little.) The stress fields around dislocations also give rise to mechanical forces on diffusing atoms. It is for this reason that precipitation takes place on dislocations at high temperature.

The most important example of diffusion in a field for semiconductors is diffusion in an electric field, since it often happens that the diffusing atoms are charged. Although it is not immediately obvious, the problem turns out to be the same as that of the moving interface, considered in Section 2.2.6. Consider what happens when a potential difference is applied to a solid containing mobile charged particles. Initially a particle accelerates away under the effect of the electrical force. After travelling a short distance it suffers a collision, loses most of the velocity it has gained, and starts to accelerate again. The result of these continual collisions is that an average 'terminal velocity' is reached which is proportional to the applied field. Mobility, B, is defined by $v = B\mathscr{E}$, where, v is the mean velocity of the particles and $\mathscr{E}$ is the applied field. It is related to diffusion coefficient by Einstein's relation:

$$\frac{D}{B} = \frac{kT}{ze} \tag{2.54}$$

where k is Boltzman's constant, e is the electron charge and ze is the change on the diffusing ion. Mobility is sometimes alternatively defined by $v = B'\mathscr{F}$ where $\mathscr{F}$ is the force acting on the particle. When the force is an electric field acting on an ion, the two mobilities are related by $B = B'ze$.

The flow of atoms across any interface is due both to diffusion and to drift under the field, i.e.

$$J = -D\frac{\partial C}{\partial x} + vC$$

assuming that the drift is in the same direction as the diffusion flow. Hence Fick's law becomes:

$$\frac{\partial C}{\partial t} = -\frac{\partial J}{\partial x} = D\frac{\partial^2 C}{\partial x^2} - v\frac{\partial C}{\partial x} \qquad (2.55)$$

This equation has been encountered before; eqn. 2.38, in the case of diffusion taking place while the surface of the specimen is moving due to crystal growth. It can be seen that the two problems are equivalent if we allow the co-ordinate axes to move at velocity v, with the electrical drift. This effectively brings the movement due to the field to a standstill, and simplifies the problem to one of simple diffusion. There is now the extra boundary condition, however, that the face of the semiconductor is moving at velocity v with respect to the axes, and this is the problem of Section 2.2.6.

Returning to the system of axes fixed to the crystal, the solution to eqn. 2.55 is given by eqn. 2.39 providing the boundary conditions are:

$C = C_0$ for $x = 0$ at all time

$C = 0$ as $x \to \infty$

$C = 0$ for $x > 0$ at $t = 0$.

i.e. the usual semi-infinite specimen with constant surface conçentration. Substituting for v in eqn. 2.39 gives the solution

$$C = \frac{C_0}{2}\left(\operatorname{erfc}\frac{x - B\mathscr{E}t}{2(Dt)^{1/2}} + \exp\frac{B\mathscr{E}x}{D}\operatorname{erfc}\frac{x + B\mathscr{E}t}{2(Dt)^{1/2}}\right) \qquad (2.56)$$

if the drift due to the field is in the same direction as the diffusion flux. For an opposing field, the sign of $\mathscr{E}$ is changed.

An interesting result from eqn. 2.56 is that if two experiments are carried out with $\mathscr{E}$ in each of the possible directions, the ratio of the diffusions is independent of time

$$\frac{C^+(x,t)}{C^-(x,t)} = \exp\frac{B\mathscr{E}x}{D} \qquad (2.57)$$

where C^+ is the field-assisted distribution and C^- is the field-retarded one. Eqn. 2.57 suggests a useful experimental technique for measuring mobility of diffusing atoms by measuring both C^+ and C^-.

2.2.9 Rate limitation on the surface concentration

In a diffusion from the vapour phase the surface concentration of solute in the semiconductor is given by the segregation coefficient between the solid and the vapour. So far it has been assumed that this surface concentration is achieved immediately at the start of the diffusion, and maintained throughout the experiment. This gives rise to the erfc solution. In nature, however, nothing happens instantaneously. Solute atoms from the vapour can only join the lattice at a rate which depends on the rate at which vapour atoms impinge on the solid, the probability of such an atom sticking and the rate at which atoms are re-evaporated. The approach to surface equilibrium therefore takes a certain time, and whether or not it produces a significant effect on the results depends on whether that time is large or small compared to the diffusion time. Surface concentration might be considered to go to equilibrium 'immediately' if the rate of exchange of atoms between surface and vapour is large compared to that between the surface and the next layer of atoms in the crystal. The surface equilibrium can then be achieved before the diffusion process gets underway in the main body of the crystal.

Let the surface concentration be $C_s(t)$ and the equilibrium surface concentration be C_0. Assuming the rate of evaporation of solute from a solid is proportional to the concentration at the surface, it is given by αC_s where α is some constant. At equilibrium this becomes αC_0 and it is exactly balanced by the rate at which atoms are joining the surface. If $C_s < C_0$, the net rate at which atoms join the surface is therefore $\alpha(C_0 - C_s)$. This causes a flow of atoms J, into the solid at $x = 0$, so

$$J = \alpha(C_0 - C_s) \quad \text{at } x = 0 \tag{2.58}$$

i.e. $$-D\frac{\partial C}{\partial x} = \alpha(C_0 - C_s) \quad \text{at } x = 0 \tag{2.59}$$

It is required, therefore, to solve the diffusion equation subject to eqn. 2.59. This is done in Carslaw and Jaeger under the thin disguise of a problem to determine the temperature distribution in a semi-infinite solid which is initially at zero temperature, and is heated by radiation from the external medium. The solution is

$$\frac{C}{C_0} = \operatorname{erfc}\frac{x}{2(Dt)^{1/2}} - \exp\left(\frac{\alpha x}{D} + \frac{\alpha^2 t}{D}\right)\operatorname{erfc}\left(\frac{x}{2(Dt)^{1/2}} + \alpha\left(\frac{t}{D}\right)^{1/2}\right) \tag{2.60}$$

By writing the term $\alpha x/D$ in eqn. 2.60 as $\alpha(t/D)^{1/2}\, x/(Dt)^{1/2}$ the equation can be expressed as a function of the dimensionless parameter $x/2(Dt)^{1/2}$. Eqn. 2.60 is plotted in this way in Fig. 2.8. The different curves are drawn for values of the time-dependant dimensionless parameter $\alpha(t/D)^{1/2}$. As would be expected the surface concentration approaches C_0 with increasing time. As t goes to infinity, the

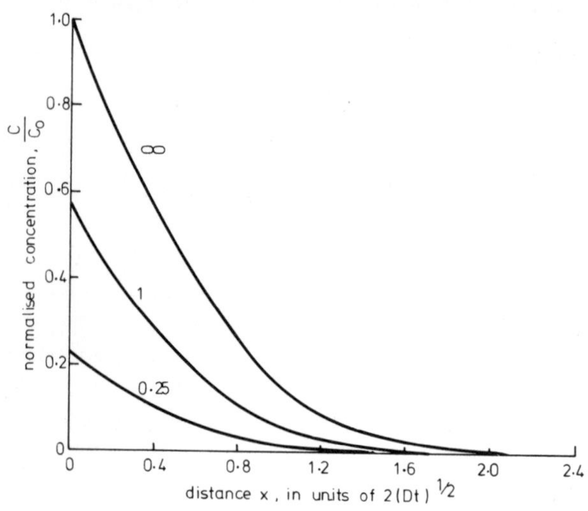

Fig. 2.8 Diffusion under conditions of a rate limitation on the surface concentration. The different profiles are plotted for increasing diffusion time using the dimensionless time-dependant parameter $\alpha(t/D)^{1/2}$

Reproduced from 'The mathematics of diffusion' by J. Crank (Clarendon Press, 1956)

concentration tends to the straightforward erfc solution. The second term becomes exp $(\alpha^2 t/D)$ erfc $(\alpha(t/D)^{1/2})$ and this approaches zero with increasing time.

The problem can be turned round to consider the out-diffusion of a doped sample into a vacuum. The solution is then:

$$\frac{C}{C_0} = \text{erf}\,\frac{x}{2(Dt)^{1/2}} + \exp\left(\frac{\alpha x}{D} + \frac{\alpha^2 t}{D}\right)\,\text{erfc}\left(\frac{x}{2(Dt)^{1/2}} + \alpha\left(\frac{t}{D}\right)^{1/2}\right) \qquad (2.61)$$

In this case, C_0 is the original doping of the crystal.

The size of any effect depends on the value of α. Work in which the out-diffusion of antimony from germanium was compared to eqn. 2.61 has been carried out by Smits and Miller.* The germanium crystals were originally homogeneously doped with antimony to a concentration of 2×10^{23} m^{-3}. The crystals were heated for three days at fixed temperatures in the range 800°C–900°C in a continuously-pumped container. Any antimony diffusing out was therefore removed. Radio-active antimony was used so that profiles could be plotted after the out-diffusion. Three curves are shown in Fig. 2.9. Curve a is the

* Smits, F. M., and Miller, R. C., *Phys. Rev.*, 1957, **107**, p. 65

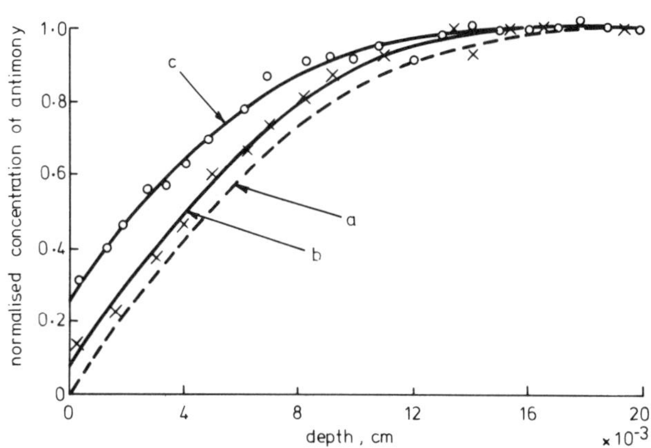

Fig. 2.9 Profiles for the out-diffusion of antimony from germanium after heating for 72 h at 850°C

Curve *a* is for $\alpha = \infty$, curve *b* is for $\alpha = 2{\cdot}4 \times 10^{-9}$ ms^{-1}, and curve *c* is for $\alpha = 3{\cdot}1 \times 10^{-10}$ ms^{-1}

Reproduced from Miller, R. C., and Smits, F. M.: *Phys. Rev.,* 1957, **107**, p. 65

erfc case, i.e. no rate limitation, $\alpha = \infty$. Curve b shows the result for the experiment described. Far from the surface going to zero immediately, it has gone to about 10 percent of the original doping in three days. The line drawn through the points comes from eqn. 2.61, putting in a value of $2{\cdot}4 \times 10^{-9}$ ms^{-1}. In a second experiment, a constriction was put into the pumping line, so that a very small vapour pressure of antimony was allowed to build up. This reduced the rate at which the solute left the solid, as is shown in curve c. The surface concentration is 30 percent of the original doping. The value of α is effectively reduced and the curve is plotted using $\alpha = 3{\cdot}1 \times 10^{-10}$ ms^{-1}.

2.3 Solutions for finite specimens

If the thickness of a sample is not large compared to $(Dt)^{1/2}$ it cannot be considered semi-infinite, and the effect of the back surface must be taken into account. The specimen may be described as being bounded by two surfaces at $x = 0$ and $x = l$.

2.3.1 Steady-state solutions

Consider a plane sheet, thickness l, with diffusion constant D, which has its surfaces maintained at C_1 and C_2 by providing different vapour pressures in the regions $x < 0$, $x > l$. After a period of time, a steady state is achieved in which diffusant enters the surface of higher concentration, C_1, say, at the same rate as it leaves the other surface. There is now no variation with time and Fick's law simplifies to:

$$\frac{d^2C}{dx^2} = 0 \tag{2.62}$$

with the boundary conditions

$$C = C_1 \text{ at } x = 0; C = C_2 \text{ at } x = l.$$

Integrating eqn. 2.62 once, we have dC/dx = constant, so the amount of diffusion flux is constant, as expected. A further integration combined with the boundary conditions gives

$$C = C_1 + (C_2 - C_1)\frac{x}{l} \tag{2.63}$$

so that the concentration varies linearly within the sheet. The flux across all sections of the sheet is given by

$$J = -D\frac{dC}{dx} = \frac{D(C_2 - C_1)}{l} \tag{2.64}$$

Eqn. 2.63 suggests a simple means of measuring the diffusion coefficient. A vapour pressure difference is maintained across a thin specimen, and the amount of solute leaving at the low concentration end is measured over a period of time. This type of experiment has been much used in diffusion work, although not with semiconductors. There are difficulties, especially with determining C_1 and C_2 at the steady state; this could presumably be overcome using radio-tracer material.

2.3.2 Diffusion from both sides of a thin specimen

There are two related problems to be considered in this category. The first concerns the indiffusion of a solute into an initially undoped specimen, with the two surfaces kept at C_0 by an external phase. In the second, outdiffusion occurs from an initially homogeneously doped sample. They are both solved in the same way, and provide interesting examples of diffusion problems that can be solved by separating the variables of distance and time. The latter problem is chosen here, as the mathematics comes out a little more easily. We start by assuming that a solution to Fick's law exists which is a product of two functions; one depending only on time, and the other only on distance, i.e.

$$C = X(x) \cdot T(t) \tag{2.65}$$

It is worth stating that the erf solutions encountered so far do not come into this class.

Substitution in the diffusion equation gives

$$X \cdot \frac{dT}{dt} = DT\frac{d^2X}{dx^2} \tag{2.66}$$

where the partial differentials have become full. Eqn. 2.66 may be written;

$$\frac{1}{DT}\frac{dT}{dt} = \frac{1}{X}\frac{d^2X}{dx^2} \qquad (2.67)$$

and in this form we have an equation which depends only on t on the left hand side, and only on x on the right hand side. The two sides can therefore be independently equated to a constant. It is convenient in the subsequent algebra to call the constant $-\lambda^2$. There are now two ordinary differential equations:

$$\frac{1}{T}\frac{dT}{dt} = -\lambda^2 D \qquad (2.68)$$

$$\frac{1}{X}\frac{d^2X}{dx^2} = -\lambda^2 \qquad (2.69)$$

The solution to eqn. 2.68 is $T = \exp(-\lambda^2 Dt)$ and that to eqn. 2.69 is $X = A \sin \lambda x + B \cos \lambda x$. Which gives for eqn. 2.67:

$$C = (A \sin \lambda x + B \cos \lambda x) \exp(-\lambda^2 Dt) \qquad (2.70)$$

Since we have put no restriction on λ, there is, in principle, an infinitely large number of solutions of the type of eqn. 2.70. The general solution is therefore

$$C = \sum_{m=1}^{\infty} (A_m \sin \lambda_m x + B \cos \lambda_m x) \exp(-\lambda_m{}^2 Dt) \qquad (2.71)$$

and the constants A_m, B_m λ_m will be determined by the initial and boundary conditions of the problem. It may be seen, by putting $t = 0$ in eqn. 2.71, that the first part of the equation is simply the Fourier series representing the initial conditions.

In the outdiffusion experiment, assume an initially homogeneously doped specimen, thickness l, concentration C_0. At the start of the experiment the surface concentrations are reduced to zero, and are kept there. The conditions are then

$$C = 0 \text{ at } x = 0 \text{ and } x = l \text{ for } t > 0 \qquad (2.72)$$

$$C = C_0 \text{ in the range } 0 < x < l \text{ at } t = 0 \qquad (2.73)$$

Condition 2.72 requires that $B_m = 0$ and $\lambda_m = m\pi/l$, where m is an integer. Combining this information with condition 2.73 gives

$$C_0 = \sum_{1}^{\infty} A_m \sin\left(\frac{m\pi x}{l}\right) \quad 0 < x < l \qquad (2.74)$$

It is now necessary to use the mathematical relationship

$$\int_0^l \sin\frac{p\pi k}{l} \sin\frac{m\pi x}{l}\, dx = \begin{cases} 0 & \text{if } m \neq p \\ \tfrac{1}{2}l & \text{if } m = p \end{cases} \qquad (2.75)$$

where m and p are both integers.

This is done by multiplying each side of eqn. 2.74 by $\sin(p\pi x/l)$ and integrating from 0 to l.

$$\int_0^l C_0 \sin\frac{p\pi x}{l}\,dx = \sum_{m=1}^{\infty}\int_0^l A_m \sin\frac{p\pi x}{l}\sin\frac{m\pi x}{l}\,dx \tag{2.76}$$

From eqn. 2.75, only one of the infinity of integrals on the right-hand side is nonzero: the one given by $p = m$. So eqn. 2.76 becomes

$$\int_0^l C_0 \sin\frac{m\pi x}{l}\,dx = A_m \cdot \frac{l}{2} \tag{2.77}$$

The left hand side of eqn. 2.77 is nonzero only for odd m. Hence

$$A_m = \frac{4C_0}{m\pi} \qquad m = 1, 3, 5 \ldots \tag{2.78}$$

The final solution is

$$C = \frac{4C_0}{\pi}\sum_{n=0}^{\infty}\frac{1}{(2n+1)}\,\frac{\sin(2n+1)\pi x}{l}\exp\left\{-\left(\frac{(2n+1)\pi}{l}\right)^2 Dt\right\} \tag{2.79}$$

where m has been replaced by $(2n + 1)$, with $n = 0, 1, 2, 3, \ldots$

In this series each term is smaller than the previous one, and the ratio between each term and the one before it gets smaller as time progresses. After a sufficiently long time, therefore, the distribution can be represented by a simple sine function in the appropriate range.

The similar problems of indiffusion from the surfaces is defined by the conditions:

$$C = C_0 \text{ at } x = 0 \text{ and } x = l \text{ for } t > 0 \tag{2.80}$$

$$C = 0 \text{ in range } 0 < x < l \text{ at } t = 0 \tag{2.81}$$

The solution may be obtained directly from eqn. 2.79 by the following argument; if we replace C by $-C$, then eqn. 2.79 represents the solution to a problem in which the slice originally has a doping level of $-C_0$, and at time $t = 0$ has its surface concentrations increased to zero. This may not mean much physically, but as far as the mathematics is concerned it is acceptable. If C_0 is added to the solution, we have a function which obeys eqn. 2.67 and conditions 2.80 and 2.81. This is therefore the solution to the indiffusion problem, and is given by

$$C = C_0\left\{1 - \frac{4}{\pi}\sum_0^{\infty}\frac{1}{2n+1}\sin\frac{(2n+1)\pi x}{l}\exp\left[-\left\{\frac{(2n+1)\pi}{l}\right\}^2 Dt\right]\right\} \tag{2.82}$$

A diagram representing eqn. 2.82 is shown in Fig. 2.10.

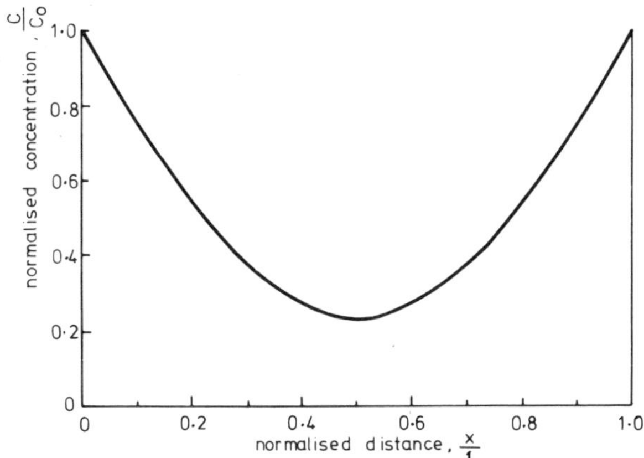

Fig. 2.10 Profile for diffusion from both sides of a thin specimen, plotted for a diffusion time $t = l^2/2D$

2.3.3 Diffusion from one side of a thin specimen

Consider the case of diffusion into a thin slab, thickness l, which has a protective coating on its surface at $x = l$, so that diffusing material enters only the face at $x = 0$. The slab is initially undoped, and the surface is maintained at C_0 throughout the experiment. The distribution of solute within the slice will, of course, depend on the boundary condition at the protected surface. Let us suppose that the protective layer has been chosen as a material in which the diffusing species is almost insoluble. The boundary condition is then that there is no flow of material from the slice into the coating, and the initial and boundary conditions for the problem become:

$$C = 0 \quad \text{for} \quad 0 < x < l \quad \text{at} \quad t = 0$$

$$C = C_0 \quad \text{at} \quad x = 0 \quad \text{for} \quad t > 0$$

$$J = -D\frac{\partial C}{\partial x} = 0 \quad \text{at} \quad x = l$$

An alternative way of looking at it is to say that since the diffusion profile cannot enter the layer, then it is reflected by the layer. The measured profile is the sum of incident and reflected atoms, and the condition $\partial C/\partial x = 0$ at $x = l$ follows.

The problem can be solved in the same way as that of Section 2.3.2, but inspection of Fig. 2.10 shows that it has, in fact, been solved already. Since the distribution in Fig. 2.10 is symmetrical about $x = \frac{1}{2}l$, it follows that $\partial C/\partial x$, at

$x = \frac{1}{2}l$, is zero. The problem is the same as that of Section 2.3.2 in all other respects. The boundary conditions correspond, and the part of the curve between $x = 0$ and $x = \frac{1}{2}l$ therefore supplies the solution. It is only necessary to note that the $\frac{1}{2}l$ of eqn. 2.82 corresponds to the l of this problem. Replacing l in eqn. 2.82 by $2l$ therefore gives the solution

$$C = C_0 \left\{ 1 - \frac{4}{\pi} \sum_0^\infty \frac{1}{2n+1} \sin \frac{(2n+1)\pi x}{2l} \exp \left[- \left\{ \frac{(2n+1)\pi}{2l} \right\}^2 \cdot Dt \right] \right\} \quad (2.83)$$

2.4 Simultaneous diffusion and chemical reaction

In this section, the case of a substance diffusing in a crystal with which it can react chemically is considered. The concentration of diffusing atoms is determined, in this case, both by the diffusion process and by the rate of loss of atoms to the chemical reaction. We assume that once an atom has been used chemically, then it is fixed in the lattice and can take no further part in the diffusion. In general, this problem can become very complex. Here we shall deal only with the simplest case, however, Crank goes into the solutions for a variety of possible situations. Mathematically the problem is closely similar to the substitutional-interstitial mechanism of diffusion, which is dealt with in a later chapter.

Assume that the rate of chemical reaction is very fast compared to the diffusion process. It is then true, to a good approximation, that at any instant, chemical equilibrium exists between the diffusing species and the immobile reactant at all points in the crystal. In the simplest possible case the concentration of reacted atoms, C_R say, is directly proportional to the diffusant concentration, i.e.

$$C_R = KC \quad (2.84)$$

Fick's law in its usual form no longer applies. The rate of increase of 'free' atoms is now equal to that due to diffusion, less the rate of loss due to the chemical reaction, i.e.

$$\frac{\partial C}{\partial t} = D \frac{\partial^2 C}{\partial x^2} - \frac{\partial C_R}{dt} \quad (2.85)$$

which, on substituting from eqn. 2.84 becomes

$$\frac{\partial C}{\partial t} = \frac{D}{1+K} \frac{\partial^2 C}{\partial x^2} \quad (2.86)$$

This is of the form of the normal diffusion equation with a revised diffusion coefficient $D/1+K$. It may be seen that if $K >> 1$, then the effect of the reaction is to considerably reduce the diffusion coefficient. Solutions for cases of diffusion with reaction are therefore identical to those for simple diffusion problems, providing the reaction can be described by an expression of the form of eqn. 2.84. In many cases eqn. 2.84 does not apply, and then eqn. 2.85 can become difficult to solve.

2.5 Effect of crystal anisotropy on diffusion

So far we have assumed D to be a constant which relates concentration gradient to flux of atoms, and is given by $\mathbf{J} = -\mathbf{D}\nabla\mathbf{C}$. According to this view the vectors of flux and gradient must lie in the same direction in space. It will be shown in this section that this is true only for diffusion in a cubic crystal. In general, D might have different values in different crystal directions, and a situation can arise in which the two vectors are not parallel. Under these circumstances, D must be raised to the dignity of a tensor.

Consider diffusion taking place in a crystal, and choose a set of orthogonal axes x, y, z, not necessarily related to the crystal axes. In the general case $\mathbf{J}$ and $\nabla\mathbf{C}$ are not parallel, and if concentration gradients exist along the three axes, then the fluxes are given by

$$\begin{aligned} J_x &= -D_{11}\frac{\partial C}{\partial x} - D_{12}\frac{\partial C}{\partial y} - D_{13}\frac{\partial C}{\partial z} \\ J_y &= -D_{21}\frac{\partial C}{\partial x} - D_{22}\frac{\partial C}{\partial y} - D_{23}\frac{\partial C}{\partial z} \\ J_z &= -D_{31}\frac{\partial C}{\partial x} - D_{32}\frac{\partial C}{\partial y} - D_{33}\frac{\partial C}{\partial z} \end{aligned} \tag{2.87}$$

The set of nine numbers D_{ij} is conveniently written

$$D_{ij} = \begin{vmatrix} D_{11} & D_{12} & D_{13} \\ D_{21} & D_{22} & D_{23} \\ D_{31} & D_{32} & D_{33} \end{vmatrix} \tag{2.88}$$

The significance of the off-diagonal elements in this matrix is that they represent a flux at right angles to the applied concentration gradient. A gradient in the x direction, $\partial C/\partial x$, now gives rise to fluxes of $-D_{11}\,\partial C/\partial x$, $-D_{21}\,\partial C/\partial x$, and $-D_{31}\,\partial C/\partial x$, in the x, y and z directions respectively.

It can be shown that, in the case of crystal systems for which the crystal axes are mutually perpendicular eqn. 2.88 can be simplified considerably. These are the cubic, tetragonal, hexagonal and orthorhombic crystal systems. If the x, y, z axes are chosen to correspond to the appropriate crystal axes, eqn. 2.88 becomes

$$D_{ij} = \begin{vmatrix} D_{11} & 0 & 0 \\ 0 & D_{22} & 0 \\ 0 & 0 & D_{33} \end{vmatrix} \quad (2.89)$$

This may be seen by the following argument; the systems mentioned above have in common the fact that three mutually perpendicular axes can be chosen with the property that a rotation of 180° about an axis brings the crystal into coincidence with itself. For the cubic, tetragonal and orthorhombic systems, these axes are the [100], [010] and [001] directions. For the hexagonal system, the first of the axes is the c axis, perpendicular to the basal plane. The second and third are any two mutually perpendicular directions in the basal plane which obey the 180° rotation stipulation.

In order to fix ideas, take the cubic crystal as an example. Impose a concentration gradient in the y direction, $\partial C/\partial y = g$. This gives rise to a flux in the x direction, $J_x = -gD_{12}$. Because of the symmetry property mentioned above, we can rotate the crystal by 180° about the x axis and restore the original situation (the concentration gradient is not to be rotated in this thought experiment). If the cartesian axes are fixed to the crystal axes, the effect of the rotation is to do nothing to the x axis, but to interchange the $+y$ and $-y$ axes. The original concentration gradient therefore becomes $-g$ and the flux in the x direction becomes $J_x = +gD_{12}$. Since the rotation restored the original situation exactly, we must come to the conclusion that the flux in the x direction is identical in both situations and $gD_{12} = -gD_{12}$. Hence $D_{12} = 0$, due to the symmetry of the crystal.

The same argument could be rehearsed for all off-diagonal elements of D_{ij}, for all systems which are invariant under 180° rotation about three crystal axes. The double-suffix rotation can now be relaxed for the diagonal elements. D_1, D_2 and D_3 are called the principal diffusion coefficients and the axes to which they refer are the principal axes. When this set is used, a gradient along the x axis produces a flux only along that same axis, although gradients which do not lie along an axis generally give rise to nonparallel fluxes.

The symmetry properties of the various crystal types can give rise to relationships between the principal coefficients. Consider a gradient $\partial C/\partial x$ causing a flux $J_x = -D_1 \, \partial C/\partial x$ in a cubic crystal. Since the x and y directions are indistinguishable, we can, if we choose, relabel the x and y axes by changing them round, so that the x axis becomes the y axis and vice versa. The situation would then be described by the relation $J_y = -D_2 \, \partial C/\partial y$. Since the relabelling has done nothing to alter the physical problem, it follows that $J_x = J_y$ and $D_1 = D_2$. This argument gives rise to the following relationships:

cubic $D_1 = D_2 = D_3$

tetragonal $D_1 = D_2 \neq D_3$

hexagonal $D_1 = D_2 \neq D_3$

orthorhombic $D_1 \neq D_2 \neq D_3$

Given the fact that flux and gradient may not be in the same direction, it is necessary to give some thought to what exactly is meant by diffusion coefficient. When the principal axes are used, the diffusion equations become

$$J_x = -D_1 \frac{\partial C}{\partial x}, \quad J_y = -D_2 \frac{\partial C}{\partial y}, \quad J_z = -D_3 \frac{\partial C}{\partial z} \tag{2.90}$$

If a gradient is imposed only along the x axis then the resulting flow will also be only along that axis, and the two vectors are parallel. Consider, however, an experiment in which a parallel-sided slab is cut from a crystal, the perpendicular to the faces of the slab being in some random direction. Impose a concentration gradient, $\nabla \mathbf{C}$ across the slab, along the perpendicular to the faces. The resulting flux of atoms will not now be parallel to the gradient. The flux measured in an experiment will be the component of the total flux resolved in the direction of the gradient. Let **i**, **j**, **k** be unit vectors along the principal axes. **J** and $\nabla \mathbf{C}$ may then be written:

$$\mathbf{J} = J_x\mathbf{i} + J_y\mathbf{j} + J_z\mathbf{k} \tag{2.91}$$

$$\nabla C = \nabla C_x\mathbf{i} + \nabla C_y\mathbf{j} + \nabla C_z\mathbf{k} \tag{2.92}$$

where $\partial C/\partial x$ has been written ∇C_x.

Substituting eqns. 2.90 into eqn. 2.91:

$$J = -(D_1 \nabla C_x\mathbf{i} + D_2 \nabla C_y\mathbf{j} + D_3 \nabla C_z\mathbf{k}) \tag{2.93}$$

The measured value of flux will be given by

$$J_m = \mathbf{J} \cdot \frac{\nabla C}{|\nabla C|} \tag{2.94}$$

where $\nabla C/|\nabla C|$ is a unit vector in the gradient direction. The measured value of diffusion coefficient will therefore be

$$D_m = -\frac{J_m}{\nabla C} = \frac{J \cdot \nabla C}{|\nabla C|^2}. \tag{2.95}$$

Combination of eqns. 2.92, 2.93 and 2.95 gives:

$$D_m = \frac{D_1 \nabla C_x^2 + D_2 \nabla C_y^2 + D_3 \nabla C_z^2}{\nabla C_x^2 + \nabla C_y^2 + \nabla C_z^2} \tag{2.96}$$

This simplifies considerably for the cases where $D_1 = D_2$:

$$D_m = \frac{D_1 (\nabla C_x^2 + \nabla C_y^2) + D_3 \nabla C_z^2}{|\nabla C|^2} \tag{2.97}$$

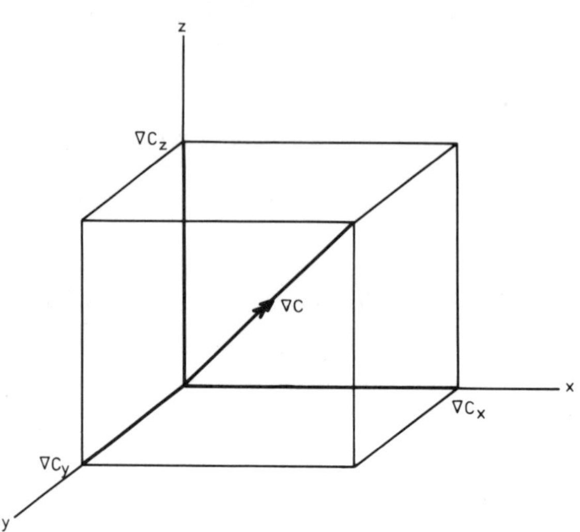

Fig. 2.11 Concentration gradient resolved along Cartesian axes

Reference to Fig. 2.11 shows that this may be written

$$D_m = D_1 \sin^2 \theta + D_3 \cos^2 \theta \qquad (2.98)$$

where θ is the angle between the direction of the imposed gradient, and the 'odd' principal axis. For cubic crystals, eqn. 2.96 becomes $D_m = D_1$ = constant, so that the measured value of diffusion coefficient is independent of direction in the crystal.

Not all semiconductors are cubic in structure, and it would therefore appear that effects of anistropy in diffusion should be important in some cases. There seems to have been very little work done, to date, investigating these effects. Some measurements have been carried out in noncubic metals (e.g. the hexagonal metals cadmium, zinc, magnesium) however, which show quite large variations of diffusion coefficient with direction.

equilibrium. Let the surroundings with which the system is in contact be at temperature T and let some irreversible process take place which involves an infinitesimal amount of heat, dQ, leaving the surroundings for the system. The change in entropy of the system is dS and that of the surroundings dS_R. The second law states that

$$dS + dS_R > 0 \tag{3.1}$$

Although the process is proceeding irreversibly, there is no reason why the heat should not be transferred from the surroundings into the system reversibly, i.e. we can write

$$dS_R = -\frac{dQ}{T} \tag{3.2}$$

hence

$$\frac{-dQ}{T} + dS > 0$$

or

$$dQ - TdS < 0 \tag{3.3}$$

During the course of the irreversible process, suppose the internal energy of the system changes by dU and the volume by dV. The quantities are related by the first law of thermodynamics

$$dQ = dU + PdV \tag{3.4}$$

where P is the pressure of the system. Substituting eqn. 3.4 in eqn. 3.3

$$dU + PdV - TdS < 0 \tag{3.5}$$

This inequality is true for any infinitesimal portion of the process. It is of interest to consider the form of eqn. 3.5, if further restrictions are made on the process. Suppose, for instance, it takes place at constant temperature and volume. The equation becomes

$$dU - TdS < 0$$

which can be written

$$d(U - TS) < 0 \text{ at constant } V, T \tag{3.6}$$

or $dF < 0$, where F is defined as the Helmholtz free energy,

$$F = U - TS \tag{3.7}$$

Eqn. 3.6 indicates that for a system kept at constant volume and temperature the Helmholtz free energy decreases during an irreversible process, becoming a minimum at the equilibrium state.

If the conditions of constant temperature and pressure are imposed, the inequality can be written

$$d(U + PV - TS) < 0 \text{ at constant } P, T \tag{3.8}$$

or $dG < 0$, where G is the Gibbs free energy of the system, defined as

$$G = U + PV - TS = H - TS \tag{3.9}$$

where H is the enthalpy of the system. Again, eqn. 3.8 means that in a system at constant temperature and pressure, the Gibbs free energy decreases until it assumes a minimum value, at which it is in equilibrium. The latter function is the more important of the two, since many more processes occur at constant temperature and pressure, than at constant temperature and volume.

The importance of eqns. 3.6 and 3.8 is that they tell us if a process such as, for instance, a chemical reaction will proceed or not. If the change in free energy is negative, then the reaction will occur spontaneously; if the change is positive, then it will not. It follows that chemical equilibrium occurs at a minimum in the appropriate free energy function, at which the possible processes proceed neither in one direction nor the other.

The reader may well at this point ask why it is necessary to derive eqns. 3.6 and 3.8 to obtain criteria as to whether a reaction will proceed, when the second law provides us with such a criterion directly in eqn. 3.1. The answer is that in order to apply eqn. 3.1, it is necessary to determine the entropy change of the surroundings as well as that of the system. This is inconvenient, to say the least. Eqns. 3.6 and 3.8 require information only about the system itself, and are therefore very much more useful in practise.

3.3 Vapour pressure and activity

A simple example of the free-energy condition, outlined above, may be seen by considering the equilibrium between a condensed phase A which may be liquid or solid, and its attendant vapour. Taking the former phase to be solid, and assuming constant temperature and pressure, the reaction may be written:

$$A(\text{solid}) \rightleftharpoons A(\text{gas}, P_A) \tag{3.10}$$

where P_A is the vapour pressure of the gas phase. Associated with the reaction, as written, is a free-energy change ΔG. If this is negative, the solid evaporates; implying that P_A is less than the equilibrium vapour pressure. If ΔG is positive, the vapour solidifies, and for the equilibrium case $\Delta G = 0$ and $P_A = (P_A)_{eq}$. In the special case of the condensed form of A being the standard state of the substance, we can write $(P_A)_{eq} = P_A°$. If the substance is not in its standard state, then its equilibrium vapour pressure $P_A \neq P_A{}^\circ$ (the 'eq' subscript can now be dropped), and the activity of the substance in the system a_A, is defined as

$$a_A = \frac{P_A}{P_A{}^\circ} \tag{3.11}$$

from which it follows that the activity of a substance in its standard state is unity. If the pure substance at the temperature of interest is a gas, then its standard state is the gas at one atmosphere pressure ($P_A{}^\circ = 1$ atm.). The activity is then numerically the same as the pressure of the gas phase measured in atmospheres (activity has no units).

Let us return to eqn. 3.10, considering it for the situation in which the reaction is in equilibrium, but in which the condensed phase of A is not necessarily its standard state. The Gibbs free energy for the vapour is given by eqn. 3.9. A small change in G may be expressed by

$$dG = dU + PdV - TdS - SdT + VdP \tag{3.12}$$

If the only work done by the vapour is work against pressure, then the first law of thermodynamics may be written

$$dQ = dU + PdV \tag{3.13}$$

and, if the change is reversible

$$dQ = TdS \tag{3.14}$$

Substituting expressions 3.13 and 3.14 into eqn. 3.12 gives

$$dG = VdP - SdT \tag{3.15}$$

Assuming the vapour is ideal,

$$PV = RT \tag{3.16}$$

where eqn. 3.16 refers to one mole of vapour (see Section 3.4), and R is the gas constant. Assuming the change took place at constant temperature, eqn. 3.15 becomes

$$dG = RT\,d\ln P \tag{3.17}$$

From eqn. 3.11

$$\ln a_A = \ln P_A - \ln P_A{}^\circ \tag{3.18}$$

so that, as $P_A{}^\circ$ is constant at constant temperature, eqn. 3.17 becomes

$$dG_A = RT\,d\ln a_A \tag{3.19}$$

If eqn. 3.19 is integrated between some arbitrary state of the substance and the standard state, then

$$G_A - G_A{}^{\circ} = RT \ln a_A \tag{3.20}$$

Here $G_A{}^{\circ}$ is the free energy of one mole of the substance A in its standard state, and G_A is the same quantity referred to some arbitrary state. Eqn. 3.20 will be used later in the chapter to determine the free energy of a substance in solution.

Eqn. 3.16 is true only for a vapour which behaves ideally. If this is not so, then a correction is required to eqn. 3.17, involving a function called the fugacity. This point will not be pursued here, however, since the solids and liquids of interest in this work normally have vapour pressures which are sufficiently low for eqn. 3.16 to be substantially correct.

Eqn. 3.15 will be used again in this work, and it is worth looking at it a little more closely. At first sight, it seems that the equation is of rather limited applicability, since it was derived using eqn. 3.14, which applied only to reversible changes. Nevertheless, eqn. 3.15 applies to all changes, as can be seen from the following argument. The thermodynamic functions involved in eqn. 3.15 are state functions, that is to say, they depend only on the state of the system, and do not depend at all on the history of the system before it reached the state. Suppose the Gibbs free energy of the system is changed infinitesimally by making small changes in pressure and temperature. Since the old and new values of the state functions depend only on the old and new states, the value of dG found from eqn. 3.15 cannot depend on the path. If it is correct for one path it must be correct for all paths. But we have proved it to be correct for one path, the reversible one. Therefore it is true for all paths, reversible or irreversible.

3.4 Partial molal quantities in solution

A solution consists of at least two components, and the problem naturally arises of what is the best way to express the amounts of material involved. The most obvious is to express the amount in terms of the weight of each component, and this is often done. It is simpler to relate the physical situation to theory, however, if the ratio of the numbers of atoms is used instead. It is more helpful, for instance, to know that a solution consisting of 28·1 g of silicon and 72·6 g of germanium is a mixture of 50 percent of each type of atom. A single atom is much too small a unit, and the unit conventionally used is the mole, which contains $6{\cdot}02 \times 10^{23}$ atoms. The number is known as Avagadro's number and gives the number of atoms in gram-atomic-weight of an element. Thus the 28·1 g of silicon contains Avagadro's number of atoms, as does the 72·6 g of germanium, since 28·1 and 72.6 are the atomic weights of the two elements. The idea can be extended to compounds so that, for instance, $6{\cdot}02 \times 10^{23}$ molecules of ZnSe weigh (65·4 + 79·0) g, where 65·4 and 79·0 are the atomic weights of zinc and selenium respectively.

The various properties of one mole of a pure substance can, at least in principle, be determined by experiment. The properties of a mole of the substance in a solution are less easy to determine, and indeed, it is a little difficult to be sure of what one means by molal properties in this situation. If two substances are mixed to form a solution, for instance, the volume of the solution is generally different from the sum of the volumes before mixing. It is fairly easy to explain this phenomenon in a qualitative way by noting that the forces of attraction between atoms in the solution must be different, on average, to the forces in the pure substances. It is not easy, however, to think of an experiment which would unambiguously sort out the fraction of the total volume occupied by each substance in the solution. The problem is circumvented by the definition of partial-molal volumes for the two substances. The foregoing is equally true of other thermodynamic properties, such as free energy and enthalpy, and partial-molal free energies and enthalpies are also defined.

Consider an amount of solution containing components 1, 2, 3 . . . , and take the volume as the property of interest. Following the notation of Darken and Gurry, we shall use the unprimed letter V to indicate the volume of one mole of solution, and the primed letter V' to indicate some arbitrary amount. Suppose there are n_1 moles of component 1, n_2 of component 2, etc. then the total volume will be related to the molal volume by

$$nV = V' \tag{3.21}$$

where $n = n_1 + n_2 + n_3 + \ldots$

The mole fraction N_1 of a component in the solution is given by

$$N_1 = \frac{n_1}{n}, \text{ and } N_1 + N_2 + N_3 + \ldots = 1.$$

Suppose we add to the solution a very small amount of component 1; dn_1, without changing the temperature and pressure, or the amounts of the other components. The volume V' will change by an amount dV'. The ratio $(\partial V'/\partial n_1)$ at constant T and P, n_2, n_3 is called the partial-molal volume of component 1. It is written $\bar{V}_1$, i.e.

$$\bar{V}_1 = \left(\frac{\partial V'}{\partial n_1}\right) \quad \text{constant } T, P, n_2, n_3 \ldots$$

It follows that if we keep the temperature and pressure constant, and vary the amounts of the components one by one, the total change in volume is given by

$$dV' = \bar{V}_1 dn_1 + \bar{V}_2 dn_2 + \bar{V}_3 dn_3 + \ldots \tag{3.22}$$

A different approach can be made as follows: the initial amount is taken to be a very large volume. $\bar{V}_1$ can then be interpreted without the use of infinitesimals as the increase in volume caused by the addition of one mole of component 1. The precise starting volume is unimportant, providing it is very large, since it is used only as a catalyst in this argument; we are principally concerned with smaller amounts to be added and removed. Add successively to this large volume n_1 moles of component 1, then n_2 of component 2 etc. and let $n_1, n_2, n_3 \ldots$ be chosen to be in the same ratios as the components in the original volume. The volume is increased by $n_1 \bar{V}_1 + n_2 \bar{V}_2 + \ldots$. The composition of the solution is unchanged, there are merely n more moles of it, where n is given by $n = n_1 + n_2 + n_3 \ldots$. The volume per mole of the solution also remains unchanged at V.

Now remove by mechanical means n moles of this solution. The reduction in volume must be

$$(n_1 + n_2 + n_3 + \ldots)V = nV$$

Since this returns us to our starting point (we have removed in one packet what we added in several smaller ones), the volume added in the first operation must be equal to that removed in the second, i.e.

$$nV = n_1 \bar{V}_1 + n_2 \bar{V}_2 + n_3 \bar{V}_3 + \ldots \tag{3.23}$$

But, $nV = V'$, the volume occupied by the n moles when they were part of the larger volume, so

$$V' = n_1 \bar{V}_1 + n_2 \bar{V}_2 + n_3 \bar{V}_3 + \ldots \tag{3.24}$$

Differentiation at constant temperature and pressure gives

$$dV' = n_1 d\bar{V}_1 + n_2 d\bar{V}_2 + \ldots + \bar{V}_1 dn_1 + \bar{V}_2 dn_2 + \ldots \tag{3.25}$$

Comparing eqn. 3.22 with eqn. 3.25 gives

$$n_1 d\bar{V}_1 + n_2 d\bar{V}_2 + \ldots = 0$$

which, on division by n becomes the more useful expression for one mole of solution

$$N_1 d\bar{V}_1 + N_2 d\bar{V}_2 + \ldots = 0 \tag{3.26}$$

Obviously, the same procedure could have been carried through for any of the other state properties of the solution at constant temperature and pressure. In particular, for a 2-component solution, the equation for Gibbs free energy becomes

$$N_1 d\bar{G}_1 + N_2 d\bar{G}_2 = 0 \tag{3.27}$$

This is often called the Gibbs-Duhem equation.

3.5 Relationships between partial molal quantities

For an arbitrary amount of solution of fixed composition, the free energy is given by

$$G' = H' - TS' \tag{3.28}$$

Differentiating this with respect to n_1 at constant temperature and pressure, keeping n_2, n_3 constant gives:

$$\left(\frac{\partial G'}{\partial n_1}\right) = \left(\frac{\partial H'}{\partial n_1}\right) - T\left(\frac{\partial S'}{\partial n_1}\right)$$

i.e.

$$\bar{G}_1 = \bar{H}_1 - T\bar{S}_1 \tag{3.29}$$

So the relation for partial-molal free energy is of the same form as eqn. 3.28, which refers to the free energy of the pure substance. It is a simple matter to demonstrate that this is generally true for partial molal quantities. The important differential relations for eqn. 3.28 are therefore also followed by eqn. 3.29, i.e.

$$\frac{\partial \bar{G}_1}{\partial T} = -\bar{S}_1 \qquad \text{constant P}, n_1, n_2 \ldots \tag{3.30}$$

$$\frac{\partial(\bar{G}_1/T)}{\partial(1/T)} = \bar{H}_1 \qquad \text{constant P}, n_1, n_2, \ldots \tag{3.31}$$

$$\frac{\partial \bar{G}_1}{\partial \mathrm{P}} = \bar{V}_1 \qquad \text{constant } T, n_1, n_2, \ldots \tag{3.32}$$

clearly, equations similar to the set 3.29 to 3.32 can be written for the other components.

3.6 Ideal solutions

When a substance is in solution it is not in its standard state, and its vapour pressure above the solution is generally less than P°. Suppose the solution is of substance A dissolved in B, and that the mole fraction of A in the solution is N_A. The vapour pressure of A above the solution must obviously vary between zero when $N_A = 0$, and the solution is pure B, and $P_A{}^\circ$ when $N_A = 1$, and the solution is pure A. It is found experimentally that for a limited number of binary solutions, P_A varies linearly between these extremes, i.e.

$$P_A = N_A P_A{}^\circ \tag{3.33}$$

Using the definition of activity, a_A, given in eqn. 3.11, the equation becomes

$$a_A = N_A \tag{3.34}$$

This is called Raoult's law, and any solution that obeys the law is called ideal. Most solutions are not ideal, but in almost all of them the law is obeyed by the solvent at very low concentrations of solute. This means that in the example given above, B would obey eqn. 3.34 in the region where N_B is slightly less than one.

An equation for the free energy of a substance not in its standard state has been derived in eqn. 3.20. Combining this with eqn. 3.34 gives the partial-molal free energy of one mole of component A in an ideal solution:

$$\bar{G}_A - G_A{}^\circ = RT \ln N_A \tag{3.35}$$

Various other thermodynamic quantities can now be calculated. Eqn. 3.31 gives the partial-molal enthalpy of a component, in terms of partial-molal free energy. Writing first the corresponding equation for the pure material in its standard state:

$$\frac{\partial(G_A{}^\circ/T)}{\partial(1/T)} = H_A{}^\circ \quad \text{at constant pressure} \tag{3.36}$$

and then subtracting eqn. 3.36 from eqn. 3.31,

$$\frac{\partial\left(\dfrac{\bar{G}_A - G_A{}^\circ}{T}\right)}{\partial(1/T)} = \bar{H}_A - H_A{}^\circ \tag{3.37}$$

Substituting from eqn. 3.35 gives

$$\frac{\partial(R \ln N_A)}{\partial(1/T)} = \bar{H}_A - H_A{}^\circ = 0 \tag{3.38}$$

Similarly,

$$\frac{\partial(R \ln N_B)}{\partial(1/T)} = \bar{H}_B - H_B{}^\circ = 0 \tag{3.39}$$

The two differentials are zero because N_A and N_B are not functions of temperature; they can be any number between zero and one. Physically, the two equations indicate that when the solution is made up from A and B there is no evolution of heat, i.e. the enthalpy of mixing $\Delta H_m = 0$. Again, this is due to there being no interaction between atoms in an ideal solution.

The free energy of mixing can be found as follows; an equation of the form of eqn. 3.24 is true for any thermodynamic property at constant temperature and pressure. The free-energy form for the solution of A and B is

$$G' = n_A\bar{G}_A + n_B\bar{G}_B$$

which on dividing by $(n_A + n_B)$ so that one mole of solution is under consideration, gives

$$G = N_A\bar{G}_A + N_B\bar{G}_B \tag{3.40}$$

The free energy of mixing is that of the solution as given by eqn. 3.40, less the free energies of the same amounts of A and B before mixing, i.e.

$$\Delta G_m = G - N_A G_A{}^\circ - N_B G_B{}^\circ \tag{3.41}$$

which, on substitution from eqn. 3.40 gives

$$\Delta G_m = N_A(\bar{G}_A - G_A{}^\circ) + N_B(\bar{G}_B - G_B{}^\circ) \tag{3.42}$$

Using eqn. 3.35, we have

$$\Delta G_m = RT(N_A \ln N_A + N_B \ln N_B) \tag{3.43}$$

At constant temperature the free energy of mixing can also be written

$$\Delta G_m = \Delta H_m - T\Delta S_m$$

which, in the special case of an ideal solution simplifies to

$$\Delta G_m = -T\Delta S_m$$

giving on substitution from eqn. 3.43 an expression for the entropy of mixing

$$\Delta S_m = -R(N_A \ln N_A + N_B \ln N_B) \tag{3.44}$$

The volume change on mixing can be found by a similar argument to that used above for ΔH_m. Subtracting from eqn. 3.32, the corresponding equation for the standard state, we have

$$\frac{\partial(\bar{G}_A - G_A{}^\circ)}{\partial P} = \bar{V}_A - V_A{}^\circ \quad \text{at constant } T \text{ and } N_B \tag{3.45}$$

Hence

$$RT\frac{\partial(\ln N_A)}{\partial P} = \bar{V}_A - V_A{}^\circ = 0 \tag{3.46}$$

Similarly,

$$\bar{V}_B - V_B{}^\circ = 0$$

Thus $\Delta V_m = 0$, so that the total volume of an ideal solution is equal to the sum of the volumes of the constituents before mixing.

3.7 Nonideal dilute solutions

Experiment shows that if the solute concentration in a solution is very small, then the partial pressure of the solute over the solution is proportional to the mole fraction of solute. Referring to the example of the previous section, where A is dissolved in B, this means that in the region of N_A close to zero and N_B close to unity,

$$P_A = kN_A \tag{3.47}$$

where k is a constant at a fixed temperature. Obviously, this is a relaxation of Raoult's law; eqn. 3.33, in which $k = P_A{}^\circ$ for all values of N_A. Eqn. 3.47 merely indicates that when the solution is dilute, the effect of the solute on the external vapour phase is additive, due to there being no interaction between solute atoms. Now, by definition of activity $P_A = a_A P_A{}^\circ$. Eqn. 3.47 therefore becomes

$$a_A = \left(\frac{k}{P_A{}^\circ}\right) N_A = \gamma_A N_A \tag{3.48}$$

Eqns. 3.47 and 3.48 are definitions of Henry's law. γ_A is the activity coefficient and, at constant temperature, is a constant in the region in which Henry's law holds. The extent to which γ_A differs from unity, indicates the extent to which the solute deviates from Raoult's law. Activity coefficients can be either greater or less than one. In practise, the dilution up to which Henry's law is obeyed varies from one solution to another. The point can be as low as a small fraction of 1 percent, or as high as a few percent of solute. The activity-coefficient concept can still be usefully employed outside the Henry's law region. In this case, it becomes a function of concentration.

So far no mention has been made of the behaviour of the solvent B, in a solution in which the solute obeys eqn. 3.47; we will now consider this point. The Gibbs-Duhem relation for the solution is

$$N_A d\bar{G}_A + N_B d\bar{G}_B = 0$$

Substituting from eqn. 3.19 yields

$$N_A d \ln a_A + N_B d \ln a_B = 0 \tag{3.49}$$

and from eqn. 3.48:

$$\ln a_A = \ln \gamma_A + \ln N_A \tag{3.50}$$

which, in the Henry's law region gives

$$d \ln a_A = d \ln N_A$$

so

$$N_A d \ln N_A + N_B d \ln a_B = 0 \tag{3.51}$$

now,

$$d \ln N_A = \frac{dN_A}{N_A}$$

and $N_A + N_B = 1$, $dN_A = -dN_B$. Eqn. 3.51 therefore becomes

$$-dN_B + N_B d \ln a_B = 0 \tag{3.52}$$

Integrating eqn. 3.52 gives

$$\ln a_B = \ln N_B + \ln I$$

where $\ln I$ is the constant of integration, so that

$$a_B = IN_B.$$

But since we are interested in the region in which N_B is close to unity, we can use the fact that when $N_B = 1$, then $a_B = 1$ as the activity of a pure substance is unity. Hence $I = 1$, and $a_B = N_B$. This is Raoult's law. So we reach the interesting conclusion that in a solution in which the solute obeys Henrys law, the solvent obeys Raoult's law. This behaviour is plotted in Fig. 3.1, in which activity is plotted against N_A. The ideal solution behaviour is indicated by the dotted lines. For low values of N_A, component A obeys Henry's law and component B obeys Raoult's law. For N_A approaching one, the solute-solvent relationship is reversed and A obeys Raoult's law and B obeys Henry's law.

The relations for the enthalpy, entropy, and free energy of mixing are no longer the same for the two components, since they obey different laws. As the solvent behaves ideally, we have

$$\bar{S}_B - S_B{}^{\circ} = \bar{G}_B - G_B{}^{\circ} = -R \ln N_B$$

$$\bar{H}_B - H_B{}^{\circ} = \bar{V}_B - V_B{}^{\circ} = 0.$$

For the solute,

$$\bar{G}_A - G_A{}^{\circ} = RT \ln \gamma_A N_A \tag{3.53}$$

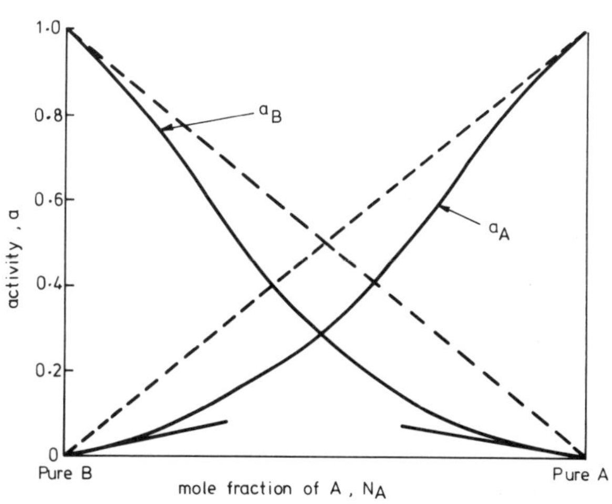

Fig. 3.1 Variation of activities in a solution of A and B. The dotted lines indicate ideal solution behaviour

The enthalpy of mixing and the volume change are given by the temperature variation of the activity coefficient

$$\bar{H}_A - H_A{}^\circ = \frac{\partial\left(\dfrac{\bar{G}_A - G_A{}^\circ}{T}\right)}{\partial\left(\dfrac{1}{T}\right)} = \frac{\partial(R \ln \gamma_A)}{\partial\left(\dfrac{1}{T}\right)} \tag{3.54}$$

This will be constant over the complete range of N_A for which Henry's law holds. A similar argument indicates that the volume of mixing is also a constant in the Henry's law region.

Writing eqn. 3.53 in the form

$$\bar{G}_A - G_A{}^\circ = RT \ln \gamma_A + RT \ln N_A$$

and comparing with eqn. 3.35, it becomes apparent that the activity coefficient of a component in solution is a measure of the deviation from ideality. An ideal solution is one in which no net interaction between atoms occurs: γ is therefore a measure of the content of the interaction in a nonideal solution. The deviation in free energy is positive for $\gamma > 1$, and negative for $\gamma < 1$.

3.8 Concentrated solutions

The previous two Sections have dealt with models for solutions that are approximations to reality at low solute levels. Concentrated solutions are a good deal more difficult to deal with; fortunately, most semiconductor solutions are quite weak, and dilute solution theory is usually adequate. In concentrated solutions there is considerable interaction between solute atoms, and the activity coefficient is a variable, as outlined above.

One further approximation to reality that is of considerable use is the regular solution. This is defined as a solution for which the partial-molal entropies are the same as for an ideal solution. It follows from this that the entropy of formation must also be the same, i.e. for a 2-component solution, eqn. 3.44 applies to a regular solution. The concept will be used later in the section on retrograde solubility.

3.9 Chemical potential (partial-molal free energy)

Consider a solution composed of n_1 moles of component 1, n_2 moles of component 2 etc. If the system is subjected to an infinitesimal change, keeping the composition constant, the variation in free energy can be expressed in terms of the changes in any two other state functions. Let us choose temperature and pressure;

$$dG' = \left(\frac{\partial G'}{\partial P}\right) dP + \left(\frac{\partial G'}{\partial T}\right) dT \tag{3.55}$$

and by comparison with eqn. 3.15 it is apparent that

$$\left(\frac{\partial G'}{\partial P}\right) = V' \quad \text{and} \quad \left(\frac{\partial G'}{\partial T}\right) = -S'$$

If the same operation is carried out allowing n_1, n_2, etc. to vary, then eqn. 3.55 becomes

$$dG' = V'dP - S'dT + \left(\frac{\partial G'}{\partial n_1}\right) dn_1 + \left(\frac{\partial G'}{\partial n_2}\right) dn_2 + \ldots \tag{3.56}$$

writing, as before, $\partial G'/\partial n_1 = \bar{G}_1$, eqn. 3.56 simplifies to

$$dG' = V'dP - S'dT + \bar{G}_1 dn_1 + \bar{G}_2 dn_2 + \ldots \tag{3.57}$$

A very similar procedure can be followed for internal energy U, expressing a small change in terms of the change in entropy, volume, and the changes in composition:

$$dU' = \left(\frac{\partial U'}{\partial S'}\right) dS' + \left(\frac{\partial U'}{\partial V'}\right) dV' + \left(\frac{\partial U'}{\partial n_1}\right) dn_1 + \ldots$$

or

$$dU' = TdS' - PdV' + \mu_1 dn_1 + \mu_2 dn_2 + \ldots \tag{3.58}$$

where $(\partial U'/\partial S') = T$, $(\partial U'/\partial V') = -P$, and $(\partial U'/\partial n_1)$ has been written μ_1. This last term is called the chemical potential, and its equivalence to partial-molal free energy will now be demonstrated.

Gibbs free energy for an indefinite amount of substance is defined as

$$G' = U' + PV' - TS'$$

which, in differential form becomes

$$dG' = dU' + PdV' + V'dP - TdS' - S'dT \tag{3.59}$$

substituting for dU' from eqn. 3.58 gives

$$dG' = V'dP - S'dT + \mu_1 dn_1 + \mu_2 dn_2 + \ldots \tag{3.60}$$

i.e.

$$\mu_1 = \left(\frac{\partial G'}{\partial n_1}\right) \quad \text{constant } T, P, n_2, n_3 \ldots \tag{3.61}$$

Physically this means that for a very large amount of solution, having free energy G', the addition of one extra mole of component 1, at some fixed temperature and pressure, will cause an increase of μ_1 in G'.

Comparing eqn. 3.61 with the notation used previously, it is apparent that there is no difference between the chemical potential μ, and the partial-molal free energy $\bar{G}$. Chemical potential is distinguished by a special name and symbol because of its considerable importance in physical chemistry (the notation will be used for the remainder of the book). Substituting for μ in eqn. 3.20 for component 1 gives the useful expression:

$$\mu_1 = G_1{}^\circ + RT \ln a_1 \tag{3.62}$$

from eqn. 3.58, μ_1 can also be written

$$\mu_1 = \left(\frac{\partial U'}{\partial n_1}\right) \quad \text{constant } S', V', n_2, n_3 \ldots \tag{3.63}$$

and it can similarly be shown that

$$\mu_1 = \left(\frac{\partial F'}{\partial n_1}\right) \quad \text{constant } V', T, n_2, n_3 \tag{3.64}$$

It is important to realise that neither eqn. 3.63 nor eqn. 3.64 represent partial-molal quantities, which are defined at constant temperature and pressure.

Consider a situation in which two phases, α and β, coexist at equilibrium and each contains the component 1. To fix ideas, component 1 might be salt, which can exist in the form of rock salt or in solution in water. An equilibrium will exist, at any given temperature and pressure, between rock salt and salt solution. Let the activities of component 1 in the phases be $a_1{}^\alpha$ and $a_1{}^\beta$ respectively. Suppose α and β exist in massive quantities, so that a single mole of 1 is an insignificant proportion of the total. The chemical potentials in the two phases are

$$\mu_1{}^\alpha = G_1{}^\circ + RT \ln a_1{}^\alpha$$
$$\mu_1{}^\beta = G_1{}^\circ + RT \ln a_1{}^\beta$$

Without changing the temperature or pressure, remove on mole of 1 from α and put it in β. This will cause a free energy change

$$\mu_1{}^\beta - \mu_1{}^\alpha = RT \ln \frac{a_1{}^\alpha}{a_1{}^\beta} \tag{3.65}$$

Now it has been pointed out in Section 3.2 that when a system is in equilibrium, it is at a minimum of free energy, so that for small deviations from equilibrium, $dG = 0$. Since α and β exist in large amounts, the transfer of a single mole of 1 is a small deviation, so the free-energy change represented by eqn. 3.65 is zero; hence

$$\mu_1{}^\alpha = \mu_1{}^\beta \tag{3.66}$$

$$a_1{}^\alpha = a_1{}^\beta \tag{3.67}$$

The result is general and leads to the important conclusion that if a component exists simultaneously in a number of phases, and if these phases are in equilibrium with each other, then the activity of the component is the same in each phase. It also leads to a conclusion that intuitively seems wrong (to the author). Since $a_1{}^{\alpha} = a_1{}^{\beta}$, the vapour pressure of the component above α is identical to that above β. This is true even if α contains 90 percent of component 1, and β contains only 1 percent. So that if, for instance, one can smell the vapour, each of the phases smells equally strongly of 1, even though one is a very strong solution, and the other is very weak.

3.10 The phase rule

This simple but important rule tells us to what extent we can fix the compositions, vapour pressures etc. of a number of phases, and to what extent these things are decided by nature.

Consider a single phase with C components, mole fractions $N_1, N_2 \ldots N_c$. The state of the phase can be completely defined by stating the temperature, pressure and the mole fractions in the phase; this amounts to (C + 2) variables. However, there is the further restriction that

$$N_1 + N_2 + \ldots + N_c = 1$$

so that if we arbitrarily fix (C + 1) of the variables, the final one is fixed by the system. This is usually expressed by saying that for a single phase there are (C + 1) 'degrees of freedom, F', i.e. F = C + 1 for a single phase.

For a system in which there are a number of phases P, say, with C components, (C – 1) mole fractions are needed to specify the composition of each phase. Therefore in order to specify the system completely, P(C – 1) + 2 data are needed, the term 2 coming from the necessity of knowing the temperature and total pressure. There are, however, further restrictions on the values of the mole fractions. At equilibrium the chemical potential of any one component must be the same in every phase, leading to the conditions

$$\mu_{11} = \mu_{12} = \ldots\ldots\ldots \mu_{1\mathrm{P}}$$

$$\mu_{21} = \mu_{22} = \ldots\ldots\ldots \mu_{2\mathrm{P}}$$

$$\vdots$$

$$\mu_{\mathrm{C}1} = \mu_{\mathrm{C}2} = \ldots\ldots\ldots \mu_{\mathrm{CP}}$$

where μ_{CP} is the chemical potential of the Cth component in the Pth phase. A total of C(P – 1) relations therefore exists between the concentrations in equilibrium, so that only

$$\mathrm{P(C-1) + 2 - C(P-1) = C - P + 2}$$

independent data are required to describe the situation completely, i.e.

$$F = C - P + 2 \tag{3.68}$$

This equation is known as the phase rule. Take as an example a 2-phase system consisting of water and its vapour. The number of components is one, water, and eqn. 3.68 gives F = 1, so that we can, within limits, determine either the temperature or the pressure. Having done this, all properties of the system are fixed. The determination can only be carried out within certain limits; if we decided to fix the temperature at 4 K, we would obviously no longer have a system containing water.

If we allow the system to contain ice as well as water and vapour, the phase rule gives F = 0, and we cannot choose any of the properties of the system. The three phases can then exist at only a single temperature and pressure: this is the well known 'triple-point' of water.

3.11 Components and phases

It is necessary at this stage to enquire a little more closely into what is meant by the terms 'component' and 'phase'. Very often the number of components is simply equal to the number of chemical elements in the system, and in the case of semiconductor systems, this is usually so. That this is not always true can be seen from the water example cited above in which water was taken as a single component, even though it consists of two elements. The number of components of a system is defined as the smallest number of constituents which can be used to describe the composition of all the phases. Since the phases could be described as H_2O (solid), H_2O (liquid) and H_2O (vapour), water served as a single component. To take another example, consider the equilibrium

$$CaCo_3 \text{ solid}) \rightleftharpoons CaO \text{ (solid)} + CO_2 \text{ (vapour)}$$

This has three elements, but can be described as

$$AB \rightleftharpoons A + B$$

where A is CaO and B is CO_2, and there are therefore only two components.

A phase is defined as a portion of a system composed of any number of chemical constituents, satisfying the requirements that it is homogeneous, and that it has a definite boundary. Two states are said to be in the same phase region, if one can be obtained from the other without discontinuities occurring in any of the state properties. Silicon containing a doping of boron atoms to a level of 10^{16} m^{-3} is therefore in the same phase region as that with 10^{17} m^{-3}; the second could be obtained from the first by a (continuous) diffusion process. Silicon liquid is in a different phase from silicon solid, however, because the liquid cannot be obtained from the solid without discontinuous changes occurring in, for instance, volume and enthalpy (the latter showing up as the latent heat of fusion).

3.12 Binary-phase diagrams

The relationships between the various phases present in a system at equilibrium are best expressed using the phase diagram for the system. These diagrams, commonly used by metallurgists, are used much less by semiconductor scientists, partly because phase diagrams for some of the most important semiconductor systems have yet to be determined. No effort will be made here to go into the subtleties of phase diagrams (Rhines, 1956),* instead, some of the major points will be considered, where they are relevant to semiconductors.

3.12.1 2-phase systems

A system containing two components can be defined by three parameters. Most usefully these are temperature, pressure and composition (only the mole fraction of one of the components is needed, since they add up to unity). A 3-dimensional representation is therefore required to show all possible conditions of the system. It is conventional to simplify this situation by taking a section through the diagram at constant pressure. This has the effect of reducing the number of axes to two so that a 2-dimensional diagram can be used. The pressure chosen, for obvious reasons, is usually one atmosphere.

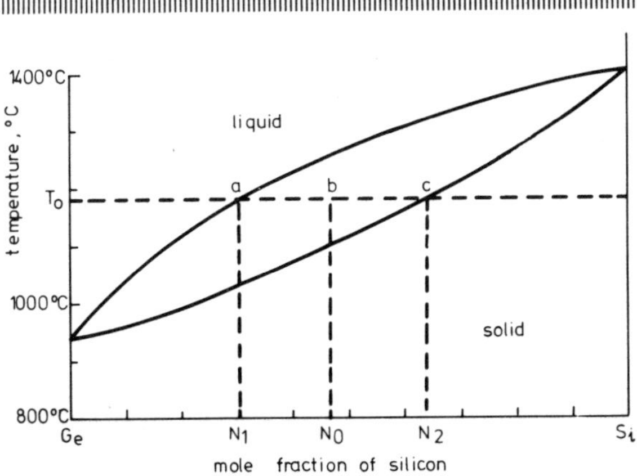

Fig. 3.2 Phase diagram for the germanium-silicon system

Reproduced from 'Constitution of binary alloys' by Max Hansen (McGraw-Hill, 1958)

The simplest type of 2-component system is that for which a solid solution exists at all concentrations: one of this type for the silicon–germanium system is shown in Fig. 3.2, in which the solid–liquid equilibrium is indicated. Fig. 3.2 shows

*Rhines, F. N.: 'Phase diagrams in metallurgy' (McGraw-Hill, 1956)

three regions; one liquid, one solid and one containing both phases. Suppose, for example, we have a solution consisting of N_1 mole fraction of silicon, and $(1 - N_1)$ of germanium. At temperatures above T_0 the solution is liquid. On reducing the temperature, it remains liquid until temperature T_0 is reached. At this temperature the upper line, called the liquidus, is arrived at, and solidification starts. The composition of the solid is given by the lower, solidus, line and is N_2 mole fraction of silicon.

From the phase rule we have two components, 2-phases and therefore two degrees of freedom. One of these has been used up in choosing the pressure of 1 atm. in the plotting of the diagram. In the region in which two phases exist (i.e. the region between the liquidus and solidus lines) there is only one degree of freedom left, so that, for instance, choosing the temperature to be T_0 fixes the equilibrium composition of the liquid at N_1 mole fraction of silicon and the composition of the solid at N_2 mole fraction. Similarly choosing either the composition of the liquid or that of the solid defines the other composition and also the temperature. Lines joining phases that coexist in equilibrium are called tielines, and it is obvious that in a 2-component system, these lines must be horizontal, since they refer to a single temperature. The line abc is thus a tieline.

Any equilibrium mixture of germanium and silicon between N_1 and N_2, at the temperature T_0, will consist of liquid given by point a on Fig. 3.2 and solid given by c. Variation within the range N_1 to N_2 changes the amounts of liquid and solid involved. This may be seen as follows. Let there be n_m moles of the mixture (Ge + Si), mean composition given by point b. Suppose that n_l moles are in the liquid phase, silicon mole fraction N_1 and that n_s moles are in the solid phase, silicon mole fraction N_2. We have

$$n_m = n_l + n_s \tag{3.69}$$

The average composition in liquid + solid is N_0 mole fraction silicon, $(1 - N_0)$ germanium. So the total number of moles of silicon is $n_m N_0$, distributed between $n_l N_1$ in the liquid and $n_s N_2$ in the solid, so

$$n_m N_0 = n_l N_1 + n_s N_2 \tag{3.70}$$

from eqn. 3.69,

$$(n_l + n_s)N_0 = n_l N_1 + n_s N_2$$

or

$$n_l(N_0 - N_1) = n_s(N_2 - N_0) \tag{3.71}$$

In terms of Fig. 3.2, this gives

$$\frac{n_l}{n_s} = \frac{bc}{ab} \tag{3.72}$$

The amount of each coexisting phase is therefore inversely proportional to the distance from the point of mean composition. This is often called the lever law and is a general result for phase diagrams. The important point to remember is that the closer the mean composition point is to the point representing the phase, the greater is the proportion of that phase.

3.12.2 3-phase systems

The reason for the simplicity of Fig. 3.2 is that germanium dissolved in silicon is in the same phase region as silicon in germanium. This, in turn, is due to the fact that these semiconductors have the same crystal structure so it is possible to go from one material to the other without any discontinuity. In general, however, in a binary system composed of materials A and B, the solid solution of A in B is a different phase to that of B in A, and there can now be three phases, namely the two solid solutions and a liquid. From the phase rule, there is only one degree of freedom in such a situation, and this is used up by plotting the diagram for a fixed pressure. Since this means that there is no choice left either for temperature or composition, the three phases can only coexist at a fixed point on the equilibrium diagram. This point is called a eutectic point, and an example of a eutectic-type system, for silicon and aluminium, is shown in Fig. 3.3. The phase diagram which, as semiconductor equilibrium diagrams go is rather simple, has a eutectic point at 577°C (called the eutectic temperature), and at about 11 percent silicon. The three phases exist together at this point, and all the other points on the diagram correspond to either one or two phases.

Two solid phases labelled α and β occur in the diagram. α exists up to 660°C and

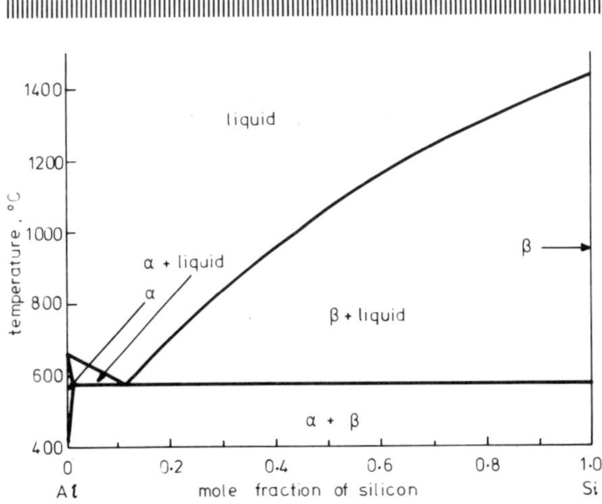

Fig. 3.3 **Phase diagram for the aluminium-silicon system**

Reproduced from 'Constitution of binary alloys' by Max Hansen (McGraw-Hill, 1958)

is a solution of silicon in aluminium: the maximum solubility is about 1·5 percent and occurs at the eutectic temperature. The line separating the α phase from the $\alpha + \beta$ region is called a solvus line. The silicon-rich side of the diagram shows no equivalent phase of aluminium dissolved in silicon; so it appears that β is pure silicon. This is not the case. A phase of aluminium dissolved in silicon does exist, but is so narrow that it cannot be shown on a diagram such as Fig. 3.3 which has a linear-composition axis. In fact the maximum solubility of aluminium is about 0·04 percent. Such a low figure might be insignificant in normal metallurgical terms, but in semiconductor terms represents quite a high doping of aluminium (about $2 \times 10^{25}\ \mathrm{m}^{-3}$). It is also of some interest that the maximum solubility does not occur at the eutectic temperature. This will be enlarged upon in Section 3.15.

It has already been noted that if equilibrium exists between two phases and the same component occurs in both, then the component will, in general, occur at different concentrations in the two phases. The ratio of the two concentrations is called the segregation coefficient (or distribution coefficient), k. The phase diagram gives a value for k between the solid and liquid phases. It is conventional to define k:

$$\mathrm{k} = \frac{\text{concentration of solute in solid}}{\text{concentration in liquid}} \tag{3.73}$$

Taking Fig. 3.2 as an example, and treating silicon as the solute, at T_0 k is given by $\mathrm{k} = N_2/N_1$. Here, $\mathrm{k} > 1$, but in a system such as that of Fig. 3.3, the solute is more soluble in the liquid phase, i.e. $\mathrm{k} < 1$. This is important in crystal growth. If, for instance, silicon crystals are grown from a melt containing a small amount of aluminium, the effect of $\mathrm{k} < 1$ is to make the growing crystal continually reject solute so that the last material to solidify is much more heavily doped than the rest.

3.13 Vapour phase

There is an apparent anomaly in the above description which the reader may well have already seen. No mention has been made of the vapour phase, even though it has been stated that any liquid or solid has, at equilibrium, a vapour phase above it. In Fig. 3.2, for instance, this would suggest that there are three phases instead of two and, according to the phase rule, that there is apparently one degree of freedom less than was assumed above. To resolve the paradox, it is necessary to consider the possible types of experimental conditions under which the equilibrium conditions for a 2-component system might be determined. There are three obvious types of experimental conditions.

(*a*) The two components are contained in an evacuated vessel. In general liquid and solid phases exist with a vapour above them. The only pressure experienced by the condensed phases, is the pressure exerted by the vapour phase. If there is only one solid, then there are three phases and one degree of freedom. This is not conventional for phase diagrams, however, because the pressure is not set at 1 atm. Pressure will, in fact, be a function of temperature and composition. Phase diagrams similar to Figs. 3.2 and 3.3 could be plotted under these conditions, and used similarly. It is still true that choosing a single variable, such as temperature, fixes the other parameters of the system, since although we have lost a degree of freedom, we have given up setting the pressure. The situation is relevant to many diffusion experiments in which a slice of semiconductor is encapsulated in a small evacuated ampoule with the doping material and then raised to a high temperature for diffusion to occur. This diffusion technique is commonly used for compound semiconductors.

(*b*) The components are kept in a cylinder by a piston which exerts a pressure of 1 atm. This is, at least in principle, the experimental condition for most phase diagrams which are plotted. It is worth considering what happens to the vapour phase. If we fix the temperature, we have used up the one available degree of freedom and the equilibrium vapour pressure is determined. Assume first of all the more common case of the equilibrium vapour pressure being less than 1 atm. The piston compresses the gas, thereby attempting to impose one more condition on the system than the phase rule will allow. The only course left open to the system is to lose a phase by condensing the vapour, leaving only liquid and solid phases, as assumed above.

If the equilibrium vapour pressure is more than 1 atm, the piston will be pushed back so as to increase the volume of the container. The liquid will start to evaporate to keep up the vapour pressure, and not until all of the material is in the vapour phase can equilibrium between vapour and piston be reached. The piston will simply increase the volume of the container to the point at which the vapour pressure is 1 atm. Phase diagrams for a fixed 1 atm pressure cannot be employed, therefore, for systems in which the vapour pressures over the condensed phases exceed 1 atm. For metals this is of academic interest only, since vapour pressures are very low, but for some compound semiconductors, very high vapour pressures can occur, and the point can become quite significant.

(*c*) The piston of (*b*) is replaced by an inert atmosphere exerting 1 atm pressure. This is a good approximation to (*b*) providing the atmosphere is truly inert and the vapour pressure is low. Almost all of the pressure on the liquid + solid system is due to the inert atmosphere, and the condensed phases can be considered separately from the vapour. The situation corresponds to the 'open-flow' type of diffusion process, in which a gas, including a vapour of the element to be diffused, is passed over the semiconductor. The process takes place in an open tube at atmospheric pressure.

It is clear that phase diagrams could be determined under any of these three conditions, and that the same is true for diffusion experiments. When using a phase diagram to help in diffusion work, it is obviously desirable for the diagram to have been plotted under similar circumstances to those of the diffusion. If it has not, then it is an approximation for that particular diffusion, although in many cases, the difference between phase diagrams plotted for the same system would be rather slight.

3.14 Activities of elements in solid solutions

As an example of the information contained in the binary-phase diagram, we now consider the variation in activity of the two elements in the region of solid solution. Fig. 3.4 shows a simple equilibrium diagram between elements A and B, containing a eutectic point. Solid solutions exist on the diagram for B dissolved in A and A in B. These phases are called α and β respectively. In the region below the eutectic temperature, and between the solvus lines, the two phases coexist, the relative amounts being given by the lever law. At a temperature T_1 the phases contain respectively $N_A{}^\alpha$ and $N_A{}^\beta$ mole fraction of the element A, and since the system is at equilibrium, the chemical potential of element A in phase α is equal to that in phase β. A similar statement can be made about element B. Both α and β are dilute solutions, so it can be assumed that Raoult's law is obeyed by the solvents, and Henry's law is followed by the solutes, i.e. in the β phase

$$\mu_B{}^\beta - G_B{}^\circ = RT \ln a_B{}^\beta = RT \ln N_B{}^\beta \tag{3.74}$$

and in the α phase,

$$\mu_B{}^\alpha - G_B{}^\circ = RT \ln a_B{}^\alpha = RT \ln N_B{}^\alpha + RT \ln \gamma_B \tag{3.75}$$

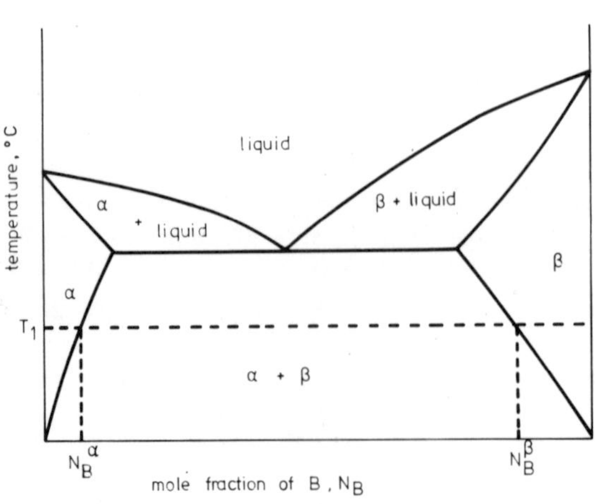

Fig. 3.4 Phase diagram showing eutectic point

since these are equal, eqns. 3.74 and 3.75 can be equated to give

$$\gamma_B = \frac{N_B{}^\beta}{N_B{}^\alpha} \tag{3.76}$$

and

$$\gamma_A = \frac{N_A{}^\alpha}{N_A{}^\beta} \tag{3.77}$$

The variations of the activities of the two elements with composition are shown in Fig. 3.5. There is no change within the solvus lines because there are no changes in the compositions of the phases: all that is changing is the amount of α relative to the amount of β. Outside the solvus lines, the activities either fall from unity with unit slope, following Raoult's law, or increase from zero with slope γ, following Henry's law.

3.15 Retrograde solubility

It has already been noted that although the solid solution phase of aluminium in silicon does exist, it is too narrow to appear on a phase diagram such as Fig. 3.3 which has a linear-composition axis. This low solubility behaviour is typical of a wide range of elements in semiconductors, and it is usual to plot the equilibrium

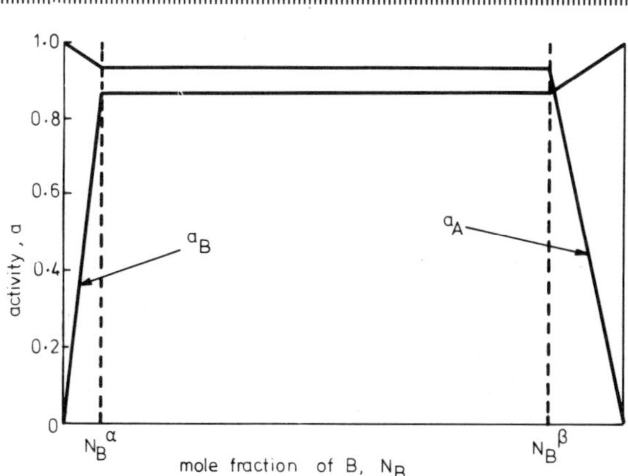

Fig. 3.5 Variation of the activities of components A and B at temperature T_1. The phase diagram for A and B is shown in Fig. 3.4

diagrams on a logarithmic scale to show the form of the liquidus and solidus curves. An example is shown in Fig. 3.6 for copper dissolved in germanium. The Figure has a eutectic temperature at about 660°C and shows the surprising result that the maximum solubility is not at the eutectic temperature but at about 880°C. This behaviour should be compared to the more conventional solubility pattern shown in Fig. 3.3 for silicon dissolved in aluminium: the maximum silicon solubility occurs at 577°C, the eutectic temperature. The phenomenon of maximum solubility occurring at a temperature higher than the eutectic temperature is called retrograde solubility, and is displayed by a wide range of elements in germanium and silicon. It has also been observed in a number of compound semiconductors.

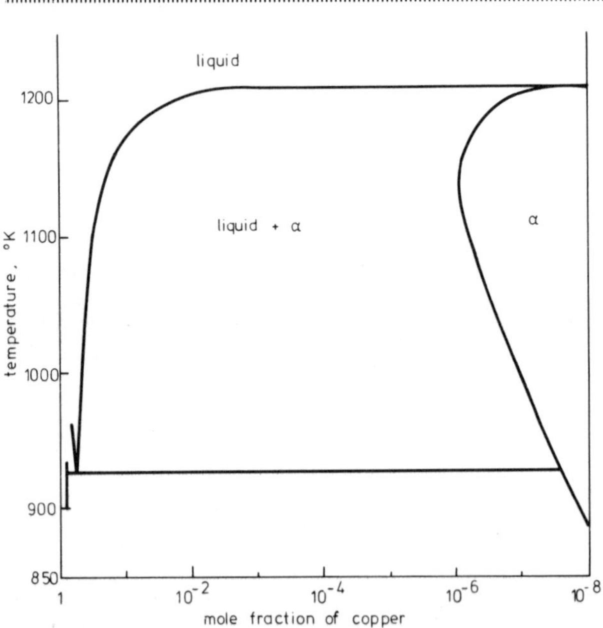

Fig. 3.6 Phase diagram for the copper-germanium system, showing retrograde solubility

Reproduced from Thurmond, C. D., and Struthers, J. D.: *J. Phys. Chem.*, 1953, **57**, p. 831

It is instructive to calculate the theoretical shape of the liquidus and solidus curves, to discover under what circumstances retrograde solubility might be expected to occur. The sections of the curves to be calculated are those between the melting point of the semiconductor and the eutectic temperature (Fig. 3.6). Let element A be the solute and B be the semiconductor. Consider the region of the diagram in which a liquid solution of elements A and B is in equilibrium with a solid solution of the same two elements. Let the mole fractions be $N_A{}^s$, $N_B{}^s$ in the solid and $N_A{}^l$, $N_B{}^l$ in the liquid. Since the solid solution is known to be dilute, $N_A{}^s$ is very small and $N_B{}^s$ is approximately unity. In order to simplify matters a little we shall assume that the liquid solution is ideal and the solid is a regular solution. The principle of the calculation is simply to note that if the solid and liquid are in equilibrium, then the chemical potential of an element in the solid must be equal to the potential of the same element in the liquid.

Consider first the equilibrium of the solute element A. The chemical potential in the solid phase, $\mu_A{}^s$, is given by eqn. 3.62.

$$\mu_A{}^s - G_A{}^\circ = RT \ln \gamma_A N_A{}^s \tag{3.78}$$

where $G_A{}^\circ$ is the free energy of one mole of pure element A in its standard state. Now the left-hand side of eqn. 3.78 represents the change in free energy in taking one mole of A from the pure state into the solid solution. It is convenient to write this term as $\Delta\bar{G}_A{}^s$ and to define in the same way $\Delta\bar{H}_A{}^s$, $\Delta\bar{S}_A{}^s$. Then

$$\mu_A{}^s - G_A{}^\circ = \Delta\bar{G}_A{}^s = \Delta\bar{H}_A{}^s - T\Delta\bar{S}_A{}^s \tag{3.79}$$

Hence

$$RT \ln \gamma_A N_A{}^s = \Delta\bar{H}_A{}^s - T\Delta\bar{S}_A{}^s \tag{3.80}$$

We now use the assumption that the solid solution is regular. By the definition of such a solution given in Section 3.8, the entropy of mixing is given by eqn. 3.44 as

$$\Delta S_m = -R(N_A \ln N_A + N_B \ln N_B)$$

But ΔS_m must also be given by

$$\Delta S_m = N_A(\bar{S}_A - \bar{S}_A{}^\circ) + N_B(\bar{S}_B - S_B{}^\circ) \tag{3.81}$$

(eqn. 3.42). Comparing eqn. 3.81 with eqn. 3.44 gives

$$\bar{S}_A - S_A{}^\circ = \Delta\bar{S}_A = -R \ln N_A \tag{3.82}$$

and

$$\bar{S}_B - S_B{}^\circ = \Delta\bar{S}_B = -R \ln N_B \tag{3.83}$$

So for the component A in the solid phase

$$\Delta\bar{S}_A{}^s = -R \ln N_A{}^s \tag{3.84}$$

Substituting eqn. 3.84 into eqn. 3.80 gives an equation for γ_A

$$RT \ln \gamma_A = \Delta\bar{H}_A{}^s \tag{3.85}$$

Hence eqn. 3.78 becomes

$$\mu_A{}^s - G_A{}^\circ = \Delta\bar{H}_A{}^s + RT \ln N_A{}^s \tag{3.86}$$

Next we look at the potential of the solute in the liquid phase, which will be treated simply as an ideal solution. There is one small complication to be taken into account. When an ideal liquid solution is made up from liquid elements A and B, no heat is evolved and this simplifies the situation. However, it is quite likely that at the temperatures of interest, both of the pure elements will be solid. This must be the case for the pure semiconductor, element B, since the temperature range of interest is that above the eutectic temperature T_e and below the melting point of the semiconductor T_B (Fig. 3.6). For the solute element there are two possible cases: (*a*) the melting point of pure A, T_A, is greater than T_B. Pure A is then always solid within the range $T_e < T < T_A$. (*b*) $T_A < T_B$ (Fig. 3.3). This makes pure A a liquid in the upper part of the range and a solid in the lower part. We shall consider the first case and look at the implications of the second at the end of the calculation.

To put it another way, we cannot make a liquid solution of A and B at temperature T, ($T_e < T < T_A$) without first turning the solids at temperature T into liquids. The free energies of fusion must therefore be involved. If the free energy of one mole of pure liquid A is g, the free-energy change in putting liquid A into the liquid solution is given by eqn. 3.35 as

$$\mu_A{}^l - g = RT \ln N_A{}^l \tag{3.87}$$

However, we must allow for the fact that the standard state of A is not pure liquid but pure solid. The free-energy change on turning one mole of solid A at temperature T into one mole of liquid is

$$g - G_A{}^\circ = \Delta G_A{}^f = \Delta H_A{}^f - T\Delta S_A{}^f \tag{3.88}$$

where $\Delta H_A{}^f$ and $\Delta S_A{}^f$ represent the changes in enthalpy and entropy occurring in the solid/liquid transition. (The reader should not be disconcerted at being asked to consider A as a liquid at a temperature below its melting point. Super-cooled liquids do exist. 'I imagine it, therefore it is', as Descartes might have said). Substituting in eqn. 3.87 for g

$$\mu_A{}^l - G_A{}^\circ = \Delta H_A{}^f - T\Delta S_A{}^f + RT \ln N_A{}^l \tag{3.89}$$

Since at equilibrium $\mu_A{}^l = \mu_A{}^s$, eqns. 3.86 and 3.89 can be combined to give

$$RT \ln \left(\frac{N_A{}^s}{N_A{}^l} \right) = \Delta H_A{}^f - \Delta \bar{H}_A{}^s - T\Delta S_A{}^f \tag{3.90}$$

where $N_A{}^s/N_A{}^l$ is the segregation coefficient.

$\Delta H_A{}^f$ can be eliminated from eqn. 3.90 by noting that at constant pressure, it is equal to the latent heat of fusion. The change of entropy at the melting point is therefore

$$\Delta S_A{}^f = \frac{\Delta H_A{}^f}{T_A} \tag{3.91}$$

Substituting in eqn. 3.90 for $\Delta H_A{}^f$ gives

$$RT \ln \frac{N_A{}^s}{N_A{}^l} = \Delta S_A{}^f(T_A - T) - \Delta \bar{H}_A{}^s \tag{3.92}$$

So far, we have only the ratio of $N_A{}^s$ and $N_A{}^l$. To find them separately, it is necessary to consider the equilibrium of the solvent element B in much the same way as has been done for the solute. Taking first of all the solid solution, we can use the fact that in a very dilute solution the solvent obeys Raoult's law

$$\mu_B{}^s - G_B{}^\circ = RT \ln N_B{}^s \tag{3.93}$$

but since $N_B{}^s$ is almost exactly unity

$$\mu_B{}^s = G_B{}^\circ \tag{3.94}$$

For the liquid solution there is an equation analogous to eqn. 3.89

$$\mu_B{}^l - G_B{}^\circ = \Delta H_B{}^f - T\Delta S_B{}^f + RT \ln N_B{}^l \tag{3.95}$$

Again putting $\mu_B{}^l = \mu_B{}^s$, at equilibrium

$$RT \ln N_B{}^l = RT \ln (1 - N_A{}^l) = -\Delta H_B{}^f + T\Delta S_B{}^f \tag{3.96}$$

An equation similar to eqn. 3.91 can again be used to eliminate a variable. This time, however, it is more convenient to eliminate $\Delta S_B{}^f$, replacing it with $\Delta H_B{}^f/T_B$. Hence eqn. 3.96 becomes

$$\ln (1 - N_A{}^l) = \frac{\Delta H_B{}^f}{R} \left(\frac{1}{T_B} - \frac{1}{T} \right) \tag{3.97}$$

which is the equation of the liquidus.

If $\Delta H_B{}^f$ is known, the liquidus can be plotted. $N_A{}^l$ can then be substituted into eqn. 3.92 and the solidus for $N_A{}^s$ can be plotted. This process has been carried out by Thurmond and Struthers (1953)*, taking germanium as the semiconductor (element B). Germanium has a melting point of 936°C and heat of fusion, $\Delta H_B{}^f$ of 8100 cal/mole: these figures substituted in eqn. 3.97 give the liquidus, shown as the full line in Fig. 3.7. The solidus can be plotted if $\Delta S_A{}^f$ and $\Delta \bar{H}_A{}^s$ are known. A reasonable average value for metals of 3 e.u. was taken for $\Delta S_A{}^f$. The dotted lines in Fig. 3.7 show solidus lines plotted from eqn. 3.92 using various values for the parameter $\Delta \bar{H}_A{}^s$ ranging from 22 k cal/mole to 5·5 k cal/mole. Retrograde solubility is shown by curves *a, b, c* and not by the curves for lower $\Delta \bar{H}_A{}^s$. It is evident that the effect is related to a large value of the parameter $\Delta \bar{H}_A{}^s$: it is not observed in most metallic systems because $\Delta \bar{H}_A{}^s$ tends to be much lower than the values used in Fig. 3.7. It can be shown that a large value of $\Delta \bar{H}_A{}^s$ is associated with systems exhibiting low solute solubilities. Most systems involving germanium and

* Thurmond, C. D., and Struthers, J. D.: *J. Phys. Chem.*, 1953, **57**, p. 831

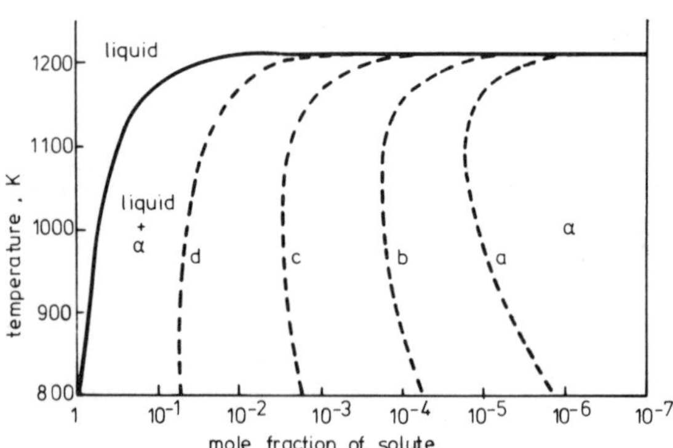

Fig. 3.7 Phase diagram for a germanium-solute system. The solidus lines *a, b, c, d* are plotted using values of $\Delta \bar{H}_A{}^S$ of 22, 16·5, 11, 5·5 k cal/mole respectively. Retrograde solubility is shown by curves a, b, c

Reproduced from Thurmond, C. D., and Struthers, J. D.: *J. Phys. Chem.,* 1953, **57**, p. 831

silicon as solvents come into this category, as do many systems with compound semiconductors.

If the melting temperature of the solute is greater than the eutectic temperature but below the melting point of the semiconductor, case (*b*) above operates. In the temperature range where the standard state of the solute is liquid, the term $\Delta H_H{}^f$ would be absent from eqn. 3.90 and the retrogression would occur for lower values of $\Delta \bar{H}_H{}^s$.

3.16 Ternary phase diagrams

Many semiconductors are compounds rather than elements, so that if a doping element is added by diffusion, or other means, the system contains three components. Such systems are called ternary systems, and are usually represented by triangular phase diagrams. An example is shown in Fig. 3.8.

To represent the phase equilibriums of a ternary four parameters are needed; temperature, pressure, and two to indicate composition. If some assumption is made about pressure, such as restricting the diagram to one atmosphere, then we are left with the necessity for a 3-dimensional diagram. The model used is a prism in which the temperature is plotted on the vertical axis and the base is used to represent composition. It is convenient, for reasons which will become apparent below, to take the base as an equilateral triangle. A constant-temperature section of the 3-dimensional diagram therefore appears as shown in Fig. 3.8. Solid and liquid

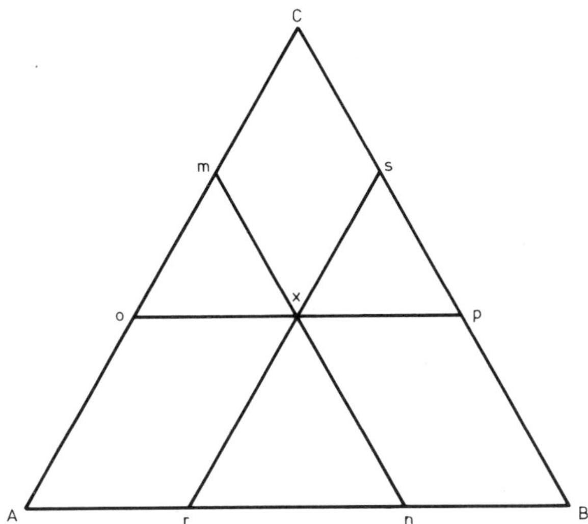

Fig. 3.8 Ternary phase diagram

phases are represented on the triangle by lines, as with the binary diagram, and phases which coexist in equilibrium are joined by tielines, as before. Tielines are drawn quite simply on binary diagrams such as Figs. 3.2 and 3.3, using the criterion that they must join points at the same temperature. All lines parallel to the horizontal axis are therefore tielines. No such simple criterion can be used for Fig. 3.8, however, since any line drawn on that Figure is a constant-temperature line. For ternaries the tielines are best determined by experiment.

The composition axes can be plotted either in terms of weight percent or mole fraction. In semiconductor work the latter is the more convenient. The three corners of the triangle represent pure A, B, C, respectively. Assuming the sides of the triangle to be of unit length, the composition corresponding to a point, such as x, within the triangle is found as follows. Lines are drawn through x parallel to the sides of the triangle. The mole fractions of the constituents are then

$$\text{mole fraction of } A = mC = nB = xs = xp$$

$$\text{mole fraction of } B = rA = sC = xo = xm$$

$$\text{mole fraction of } C = oA = pB = xr = xn$$

where the property of an equilateral triangle of unit side is used that

$$xs + xo + xr = 1.$$

Hence the point x corresponds to the composition $0{\cdot}3A$, $0{\cdot}3B$, $0{\cdot}4C$.

Fig. 3.9 shows an example of a ternary phase diagram involving the semiconductor GaAs (Panish, 1966).* The system is for Ga – As – Zn at 1050°C, and is one of the most carefully investigated compound semiconductor systems both from the point of view of solubility of zinc in GaAs and from that of diffusion. Zinc, being from group II of the periodic table, is an acceptor in GaAs and is the most commonly used element for preparing *p*-type material. Three distinct regions exist on Fig. 3.9: a liquid phase with a liquidus line delineating its boundary, a region in which solid zinc doped GaAs is in equilibrium with the liquid, and the solid GaAs phase. Again the diagram has been chosen for its relative simplicity. The system becomes a little more complicated at lower temperatures, when further solid phases such as Zn_3As_2 appear.

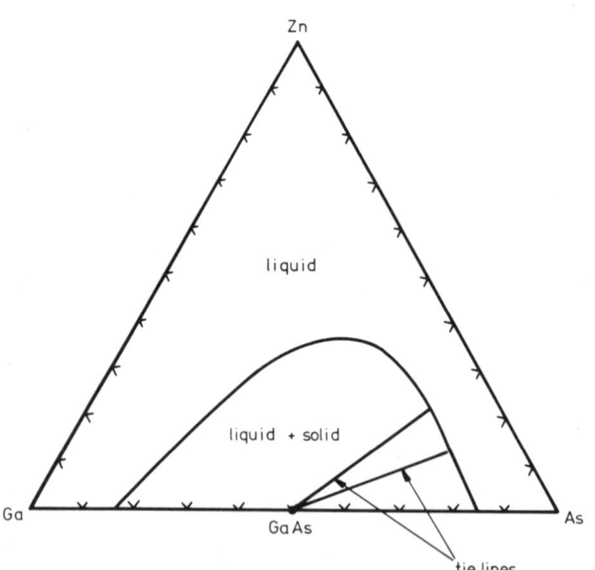

Fig. 3.9 Phase diagram for the gallium-arsenic-zinc system at 1050°C

Reproduced from Panish, M. B.: *J. Phys. & Chem. Solids,* 1966, **27**, p. 291

The solid phase appears as just a point on Fig. 3.9 because of the small extent of the phase (a similar effect was noted in connection with Fig. 3.3). If the region close to GaAs were expanded it would appear as an area of solid solution bounded by a solidus line: in fact the greatest solubility of zinc in the solid solution is about

* Panish, M. B., *J. Phys. & Chem. Solids*, 1966, **27**, p. 291

0·006 mole fraction. Tielines join points on the solidus to the corresponding equilibrium points on the liquidus. It is not obvious at which point on the solidus they should start, but since the solidus is so small, it does not really matter. In the region for which both phases exist the relative amounts of each phase at equilibrium are given by a lever law as described for the binary diagram. It is worth restating the fact that although the extent of the solid solution is trivially small in chemical terms, the amount of zinc in the GaAs has a drastic effect on the electrical properties of the semiconductor.

3.17 Case study: diffusion of zinc into GaAs

In this section, an effort will be made to demonstrate the way in which knowledge of the appropriate phase diagram can be of help in understanding a diffusion situation. The zinc/GaAs system has been chosen because of the large amount of data available for it. Diffusion into compound semiconductors is usually carried out in closed rather than open-flow systems, because of the high vapour pressures of many of these materials at diffusion temperatures. GaAs, for instance, has a high overpressure of arsenic at elevated temperatures, and an open-flow system containing the diffusing species only (zinc) would continually remove arsenic from the vicinity of the semiconductor, which would replace it by continuing to evaporate. Obviously, over a period of time this process would seriously deplete the semiconductor so that eventually only pure gallium would be left. To prevent this happening one could arrange to have just the right vapour pressure of arsenic in the flowing gas, as well as zinc. In principle, an equilibrium situation could then be achieved, but it is rather difficult to do in practise, and it is much simpler to carry out the diffusion in a closed evacuated ampoule.

The experimental technique is to put a suitably prepared slice of GaAs into a silica ampoule, together with a weighed amount of zinc and (probably) a weighed amount of arsenic. The ampoule is evacuated and sealed off before being placed into a furnace at diffusion temperature. A temperature of 1050°C will be chosen so that Fig. 3.9 is the relevant phase diagram. Care must obviously be exercised in using an equilibrium diagram to describe diffusion, which is by its nature a nonequilibrium process (equilibrium is not reached until the semiconductor is uniformly doped, i.e. until the diffusion has finished). The assumption is made that the surface of the semiconductor is at equilibrium with the surrounding vapour throughout the experiment. This amounts to saying that the surface concentration goes to the equilibrium solubility at the start of the experiment and stays there. There is good evidence to show that this does, in fact, occur. Under these circumstances, the equilibrium diagram is a very good approximation to the experimental conditions, providing the amount of zinc that becomes incorporated in the semiconductor during the course of the experiment is small compared to the total amount. Strictly, the liquid + vapour part of the system is continually losing zinc owing to the diffusion process, and the equilibrium is therefore changing throughout the experiment. There is experimental evidence to show that over a wide range of diffusion conditions, this effect is small for Zn/GaAs.

The information immediately available to the experimenter comprises the weight of the GaAs slice, the weights of the added zinc and arsenic and the volume of the diffusion ampoule. In addition he chooses the diffusion temperature, 1050°C here. To make any general statement about the diffusion conditions, or to come to any conclusions about the diffusion mechanism, it is necessary to describe the diffusion conditions in terms of the gallium, arsenic and zinc vapour pressures in the ampoule, together with the composition of any liquid phase. A good deal of information on the system is required to do this. Not only is the information of the liquidus needed (Fig. 3.9) but it is also necessary to know the vapour pressures of the three elements corresponding to each point on the liquidus. Fortunately, this information has been gathered by Shih, Allen and Pearson (1968)* at Stanford, in a fine series of experiments. In Fig. 3.10 is shown the way in which the arsenic vapour pressure varies as a function of the mole fraction of arsenic in the liquid.

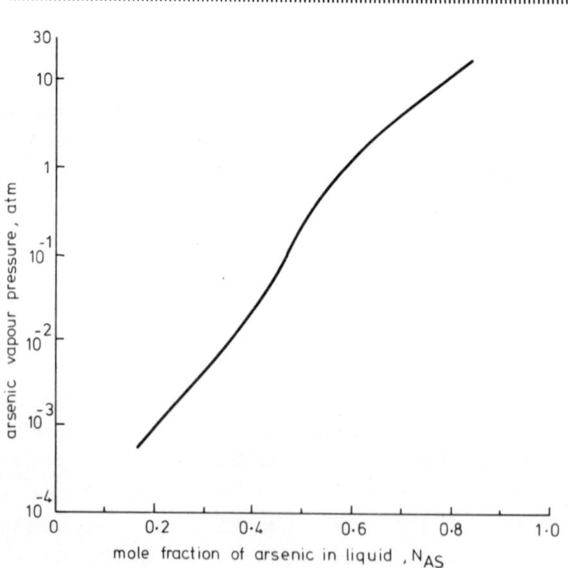

Fig. 3.10 Variation of arsenic vapour pressure with the composition of the liquid with which it is in equilibrium. Graph plotted for 1050°C

Reproduced from Shih, K. K., Allen, J. W., and Pearson, G. L.: *J. Phys. & Chem. Solids*, 1968, **29**, p. 367

* Shih, K. K., Allen, J. W., and Pearson, G. L., *J. Phys. & Chem. Solids*, 1968, **29**, p. 367

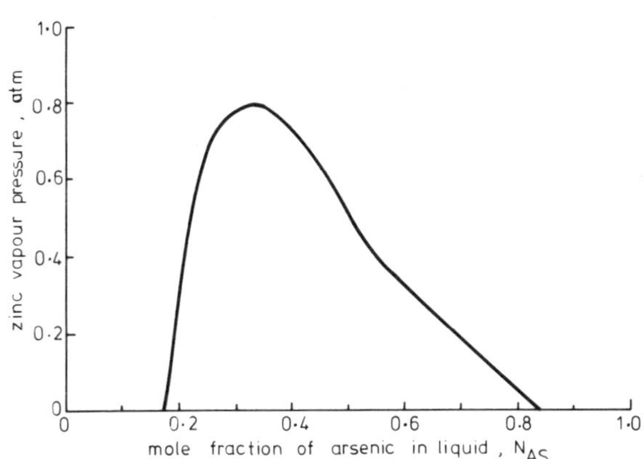

Fig. 3.11 Variation of zinc vapour pressure with the composition of the liquid with which it is in equilibrium. Graph plotted for 1050°C

Reproduced from Shih, K. K., Allen, J. W., and Pearson, G. L.: *J. Phys. & Chem. Solids,* 1968, **29,** p. 367

(Fig. 3.9 shows that a given mole fraction of arsenic in the liquid defines a unique point on the liquidus). Fig. 3.11 gives the same information for the zinc vapour pressure at 1050°C. The gallium vapour pressure is so small over the entire range that it can safely be ignored. Only the vapour pressure of the As_4 species is considered in Fig. 3.10: As and As_2 also exist in the vapour, but only in negligibly small amounts. The vapour pressures are determined from the known weights of the elements by an iterative procedure using an equation for the arsenic vapour pressure which must now be derived.

3.17.1 Arsenic vapour pressure

The pressure of As_4 in the ampoule is given by the ideal gas law

$$P_{As_4} V = nRT \tag{3.98}$$

where V is the volume of the ampoule and n is the number of moles of As_4 vapour in the ampoule. The problem is therefore to determine n in terms of the amounts of material in the ampoule.

The ampoule will contain a slice of GaAs and a small weight of added zinc, $W_{Zn}{}^a$, say. A small piece of arsenic is commonly added to the tube also. Let this weight be $W_{As}{}^a$. At diffusion temperature some of the GaAs disproportionates. GaAs is always very close to stoichiometry, so the number of moles of gallium leaving the solid is equal to the number of moles of arsenic leaving. The vapour pressure of gallium is negligibly small, so to a good approximation it can be assumed that all of the gallium leaving the GaAs goes into the liquid phase. If there are T moles in the liquid and the mole fraction of gallium is N_{Ga}, then this amounts to $N_{Ga}T$ moles which have left the solid and joined the liquid phase.

Of the arsenic that leaves the GaAs (also $N_{Ga}T$ moles), some goes into the liquid phase and some into the vapour phase. The amount of arsenic in these two phases is supplemented by the $W_{As}{}^{a}$ which has been added, amounting to an extra $W_{As}{}^{a}/M_{As}$ moles, where M_{As} is the atomic weight. The total amount of arsenic in the liquid + vapour phases combined is therefore

$$\frac{W_{As}{}^{a}}{M_{As}} + N_{Ga}T \quad \text{moles}$$

Of this, the amount in the liquid phase is $N_{As}T$ moles, where N_{As} is the mole fraction in the liquid. So the number of moles of As in the vapour would be

$$\frac{W_{As}{}^{a}}{M_{As}} + N_{Ga}T - N_{As}T \quad \text{moles} \tag{3.99}$$

But we know that the predominant vapour form is not As but As_4. It takes four moles of the former to make one of the latter, so the number of moles of As_4 in the vapour phase is

$$n = \frac{1}{4}\left(\frac{W_{As}{}^{a}}{M_{As}} + T(N_{Ga} - N_{As})\right) \quad \text{moles} \tag{3.100}$$

It remains to obtain an expression for T. This can be done in terms of the known added weight of zinc, $W_{Zn}{}^{a}$. Because of the low solubility of zinc in GaAs shown in the phase diagram, we can ignore the zinc in the solid phase. The weight of zinc in the liquid is given by $(W_{Zn}{}^{a} - W_{Zn}{}^{v})$ where $W_{Zn}{}^{v}$ is the weight in the vapour. The number of moles of zinc in the liquid is found by dividing this weight by the atomic weight of zinc, M_{Zn}. The mole fraction of zinc in the liquid can then be written

$$N_{Zn} = \frac{W_{Zn}{}^{a} - W_{Zn}{}^{v}}{M_{Zn}T}$$

or

$$T = \frac{W_{Zn}{}^{a} - W_{Zn}{}^{v}}{N_{Zn}M_{Zn}} \tag{3.101}$$

Substituting eqns. 3.101 and 3.100 in eqn. 3.98 gives

$$P_{As_4} = \frac{RT}{4V}\left\{\frac{W_{As}{}^{a}}{M_{As}} + \frac{W_{Zn}{}^{a} - W_{Zn}{}^{v}}{N_{Zn}M_{Zn}}(N_{Ga} - N_{As})\right\} \tag{3.102}$$

which is the required result.

The arsenic and zinc vapour pressures and the composition of the liquid can be found by an iterative procedure (Shih, Allen and Pearson, 1968)* using eqn. 3.102. First a guess is made as to where in Fig. 3.9 the system lies. Values of N_{Ga}, N_{As}, N_{Zn} are read from the liquidus. Fig. 3.10 gives the zinc vapour pressure for this liquid composition, and this information, together with the known volume of the ampoule allow $W_{Zn}{}^{v}$, the weight of zinc in the vapour, to be calculated. All parameters of eqn. 3.102 now have values and the partial pressure of arsenic in the ampoule can be calculated.

A second value for P_{As_4} can be obtained from Fig. 3.11, using N_{As}. The two values for P_{As_4} coincide only if the original guess was correct. A more informed guess is then made, and the procedure is repeated. Usually about three cycles gives a consistent result, and the vapour pressures of arsenic and zinc and the composition of the liquid are then known for the experimental conditions.

3.17.2 Number of degrees of freedom

Having seen how to calculate the vapour pressures and liquid composition for the experiment, it is of interest to consider how many of these parameters are under the control of the experimenter. There are three components and three phases. The phase rule

$$F = C - P + 2$$

therefore indicates two degrees of freedom. One of these was used in deciding to carry out the diffusion at 1050°C. The remaining one can be used by fixing one of the vapour pressures or the composition of the liquid. Once this is done, say by setting the arsenic pressure to 10^{-2} atm, the system is completely defined. Thus although it might seem highly desirable to carry out a set of diffusions at 1050°C in which the arsenic vapour is kept rigidly at 10^{-2} atm and the zinc pressure is varied over a range, nature will not allow it.

Under some circumstances, however, nature relents and gives us one more degree of freedom. If arsenic has been added to the ampoule, then we are dealing with the arsenic-rich side of the liquidus curve in Fig. 3.9. From Fig. 3.11 it may be seen that the arsenic vapour pressure is quite high under these circumstances. A reasonably large mass of arsenic must therefore exist in the vapour phase, unless the ampoule has a very small volume. A stage may be reached in which it is impossible to achieve the arsenic vapour pressure predicted by Fig. 3.11, even if all the added arsenic were to go into the vapour phase. Under these circumstances, a liquid phase cannot appear: if any did form, momentarily, it would immediately boil away. The number of phases is then reduced to two, and the number of degrees of freedom increased to three. This effect is really quite subtle. It depends on the weight of arsenic in the vapour phase which, in turn, depends on the ampoule volume. So it is possible, for instance, that certain amounts of semiconductor, zinc and arsenic

* Shih, K. K., Allen, J. W., and Pearson, G. L., *J. Phys. & Chem. Solids*, 1968, **29**, p. 379

would form three phases if the diffusion were carried out in a 1 ml ampoule, but only two if a 10 ml vessel were used. The number of degrees of freedom, and therefore the variables available for manipulation by the experimenter, would be different in the two cases.

3.18 Mass-action law

The law of mass action occupies an important place in the chemistry of solutions. In recent years it has been used to describe the behaviour of defects in solids, and it will be used in this way in the following chapter.

Consider a general chemical reaction at constant temperature and pressure

$$a\mathrm{A} + b\mathrm{B} \rightleftharpoons c\mathrm{C} + d\mathrm{D} \tag{3.103}$$

where A, B, C, D are symbols for the elements or species involved and a, b, c, d are the numbers of moles used up. The reaction may be between pure substances in their standard states but, in general, this will not be so. Any of the species may, for instance, appear in solution in any one of many possible solvents. The free-energy change for the reaction ΔG is simply the free energy of the products minus that of the reactants

$$\Delta G = c\bar{G}_c + d\bar{G}_D - a\bar{G}_A - b\bar{G}_B \tag{3.104}$$

Now consider the special case in which the substances are all in their standard states. Eq. 3.104 becomes

$$\Delta G^\circ = cG_c{}^\circ + dG_D{}^\circ - aG_A{}^\circ - bG_B{}^\circ \tag{3.105}$$

Subtracting

$$\Delta G - \Delta G^\circ = c(\bar{G}_c - G_c{}^\circ) + d(\bar{G}_D - G_D{}^\circ) - a(\bar{G}_A - G_A{}^\circ) - b(\bar{G}_B - G_B{}^\circ) \tag{3.106}$$

Substituting from eqn. 3.20, we have

$$\Delta G - \Delta G^\circ = cRT \ln a_c + dRT \ln a_D - aRT \ln a_A - bRT \ln a_B \tag{3.107}$$

i.e.

$$\Delta G - \Delta G^\circ = RT \ln \frac{a_C{}^c a_D{}^d}{a_A{}^a a_B{}^b}$$

Let us now take the equilibrium case, in which all four species are in equilibrium with each other. ΔG is now zero and eqn. 3.107 becomes

$$\Delta G^\circ = -RT \ln K \tag{3.108}$$

where

$$K = \frac{a_C{}^c a_D{}^d}{a_A{}^a a_B{}^b} \tag{3.109}$$

and it must be emphasised that in eqn. 3.109 all the activities refer to the equilibrium case. K is called the equilibrium constant. Since ΔG° is a function only of temperature, it follows that K also is a function only of temperature. Eqn. 3.109 is usually called the law of mass action.

As a simple example of the use of the law, let us return to the case study of Section 3.17. In that Section it was noted that the arsenic vapour in equilibrium with GaAs is virtually all As_4 species. There is also a vapour of gallium, although it was neglected in Section 3.17. The system consists of solid, liquid and vapour phases. A dynamic equilibrium must exist between the solid and vapour phases, with gallium and arsenic atoms both leaving and joining the solid surface at an equal rate. The equilibrium can be written

$$4\,\mathrm{Ga}\,(\mathrm{gas}) + \mathrm{As_4}\,(\mathrm{gas}) \rightleftharpoons 4\,\mathrm{GaAs}\,(\mathrm{solid})$$

The activities are as follows:

$$a_{Ga} = \frac{P_{Ga}}{P^\circ_{Ga}}, \qquad a_{As} = \frac{P_{As_4}}{P^\circ_{As_4}}, \qquad a_{GaAs} = 1$$

where P°_{Ga} and $P^\circ_{As_4}$ are the vapour pressures over pure gallium and arsenic at the equilibrium temperature, and the activity of a pure substance in its standard state is unity. Hence the law of mass action gives

$$\left(\frac{P_{Ga}}{P^\circ_{Ga}}\right)^4 \left(\frac{P_{As_4}}{P^\circ_{As_4}}\right) = K_1(T)$$

and, since the denominators are constants at constant temperature, we have

$$P_{Ga} = K_2(T) P_{As_4}^{-1/4}$$

so that if, for instance, the experimental conditions are changed sufficiently to increase the arsenic pressure by a factor of 16 the gallium vapour pressure is halved.

Chapter 4

Point defects in semiconductors

Diffusion in solids works on the marginal properties of the crystals in the sense that although almost every lattice site is occupied by the 'correct' atom, it is the minute fraction that are unoccupied, or are occupied by foreign atoms, that is important. Many crystal properties other than diffusion also depend in a critical way on the relatively small deviations from perfection shown by crystals, the most striking example being the way in which the electrical properties of semiconductors depend on the presence of a small proportion of doping atoms. In this chapter, point defects in semiconductor crystals are considered, and it is shown that many of the ideas outlined in the previous chapter on solutions can be carried over into the study of defects in crystals.

Imperfections in crystals can be divided into three main types, classified according to the number of dimensions taken up, i.e. planar, line, and point defects. The first are grain boundaries within the crystal: they have a considerable effect on the diffusion properties of materials, but will not be considered because virtually all semiconductor devices are made from high quality single-crystal material in which grain boundaries do not appear. Dislocations constitute the line defects of a crystal: the interaction between the diffusion properties of a crystal and the dislocation content will be described in a later chapter. Here we are concerned only with point defects. Initially elemental semiconductors are described, after which the extra complications which can come with compounds will be considered.

Let us start by conjuring up a mental picture of a perfect elemental semiconductor crystal, with the right atoms in all the right places. The ways in which deviations from perfection can occur are:

(i) atoms missing from lattice sites, i.e. vacancies
(ii) atoms present at interstitial sites, sites, which according to the crystal structure, should not be occupied at all
(iii) foreign atoms, either substituting for normal lattice sites or occupying interstitial sites
(iv) clusters, in which two or more of these point defects occur together

(v) electronic defects. The basic electronic defects are 'free' electrons in the conduction-band of the semiconductor and positive holes in the valence-band. In addition, any of the other defects listed may act as donors or acceptors, giving levels in the forbidden gap of the semiconductor and becoming charged in the process.

The analogy between the solution of a defect in a crystal and a conventional dilute solution of one material in another is fairly obvious. When the defect in question is a foreign atom, the distinction disappears altogether: the combination of defect and crystal is a solid solution in the conventional sense. The formation of any defect in a crystal must involve a change in enthalpy and entropy of the crystal, and a free energy of formation can therefore be assigned to the defect. It might be expected that at sufficiently low concentrations, Henry's law would be obeyed by the defects and this is in fact so. It follows that most of the theory of the previous chapter can be carried over for the consideration of defects in crystals. A chemical potential function can be assigned, for instance, to vacancies, and, within certain limits, the law of mass action can be used when quasichemical reactions take place between defects. All this provides a powerful technique for analysing defect interactions in solids.

4.1 The stable crystal

Having imagined a perfect crystal, it is interesting to hold on to the idea for long enough to consider whether or not such a crystal would be thermodynamically stable. The point is of some importance. Most crystals contain large quantities of dislocations, for instance, but it can be shown by thermodynamic arguments that these defects are unstable. It was no doubt this knowledge that sustained crystal growers in their efforts to grow dislocation-free semiconductor crystals. Their efforts were eventually sucessful, and some semiconductors can now be obtained virtually dislocation-free (the crystal requires very careful handling if it is to be kept in this state, of course). Let us consider an elemental crystal containing vacancies, and start with the entropy of the crystal.

So far we have taken a standard thermodynamic approach to entropy, defining it in terms of other thermodynamic functions. There is a completely different approach, by way of statistical mechanics, which in many respects is more down to earth. It amounts to saying that entropy is a measure of the disorder in a system. A given system is usually characterised by its macroscopic properties; weight, volume, pressure etc. It is a fact, however, that a given set of macroscopic parameters must correspond to a great many different configurations of the system when it is viewed on a microscopic scale. Take the example of a solid solution composed of 10 percent of substance A in 90 percent of substance B. The macroscopic statement about the composition is the single fact just stated: one atom in ten is an A atom. On a microscopic scale, however, the system would need to be characterised by stating which type of atom, A or B, occupied each individual lattice site. It is clear that the macroscopic state corresponds to many possible microscopic states, and this will be true for almost any system.

Let the number of microscopic states be $\mathscr{W}$. The entropy of the system is then defined by the Boltzman equation as

$$S = k \ln \mathscr{W} \tag{4.1}$$

where k is Boltzman's constant. If the solution of A in B is ideal, so that there is no interaction between atoms, the entropy of the solution can therefore be found very easily. We need only to calculate the number of microscopic configurations corresponding to the given composition.

Now return to the original example of a solution not of A in B, but of vacancies in an elemental crystal. Initially the crystal is perfect, with one atom corresponding to each lattice site. The number of microscopic configurations is therefore unity, i.e. $\mathscr{W} = 1$. Introduce n_v vacancies into the crystal so that there are now n_A atoms. Let us first of all ignore any interactions between vacancies and atoms, and consider the purely geometrical effect of mixing n_v vacancies and n_A atoms on $n_A + n_v$ sites. The number of ways of mixing is given by

$$\mathscr{W} = \frac{(n_v + n_A)!}{n_v! n_A!}$$

From eqn. 4.1, the entropy of mixing is

$$\Delta S_M = k \ln \frac{(n_v + n_A)!}{n_v! n_A!} - k \ln 1 \tag{4.2}$$

where the second term on the right-hand side refers to the perfect crystal and is zero. Logarithms of factorial numbers are best dealt with using Stirling's approximation:

$$\ln n! = n \ln n - n \tag{4.3}$$

The approximation is good for large values of n, and is therefore suitable. Substituting eqn. 4.3 in eqn. 4.2 yields

$$\Delta S_M = -k \left(n_v \ln \frac{n_v}{n_v + n_A} + n_A \ln \frac{n_A}{n_v + n_A} \right) \tag{4.4}$$

Now it may be seen that $n_v/(n_v + n_A)$ and $n_A/(n_v + n_A)$ are the mole fractions of vacancies and atoms, respectively, N_v and N_A say. If, in addition, we choose sufficient material for $n_v + n_A$ to add up to Avagadro's number A we can write

$$\Delta S_m = -Ak(N_v \ln N_v + N_A \ln N_A) \tag{4.5}$$

But Boltzman's constant and the gas constant R are related by $Ak = R$. Substituting this in eqn. 4.5 gives eqn. 3.44, which is the entropy of mixing of an ideal solution derived from purely thermodynamic arguments.

Hence the thermodynamic and statistical methods give the same result. Note also that the assumption used in deriving eqn. 3.44 and eqn. 4.5 was the same, namely, the ideality of the solution. It was pointed out in Section 3.7 that as soon as interaction between atoms takes place, eqn. 4.5 is no longer correct. This can be allowed for by defining an excess entropy of mixing for vacancies, ΔS_v say, that is the difference between eqn. 4.5 and the experimentally determined value, i.e.

$$\Delta S_{\mathrm{exp}} = \Delta S_m + N_v \Delta S_v$$

As defined above, the quantity ΔS_v refers to the excess entropy of one mole of vacancies. Physically ΔS_v is the result of the interaction between a vacancy and its nearest neighbours. The vibrations of the atoms in the vicinity of a certain lattice site are obviously disturbed if an atom on the site is replaced by a vacancy. The amount of disorder in the crystal is increased, giving rise to extra entropy.

The free energy of mixing can now be written as

$$\Delta G = N_v \Delta H_v + RT(N_v \ln N_v + N_A \ln N_A) - N_v T \Delta S_v \tag{4.6}$$

where ΔH_v is the enthalpy of formation for one mole of vacancies, and refers to the enthalpy required to take a vacancy from the surface of the crystal to a site within it. This amounts to defining the defect at the surface as the standard state for a vacancy. Because of the low concentration of vacancies, the second term in the mixing entropy, $N_A \ln N_A$, is approximately zero and can be ignored. The free-energy change per mole of added vacancies can therefore be written

$$\frac{\Delta G}{N_v} = \Delta H_v - T\Delta S_v + RT \ln N_v \tag{4.7}$$

The equilibrium concentration of vacancies can readily be calculated, since at that concentration $\partial G/\partial n_v$ equals zero. From eqn. 4.7 we can write

$$\frac{\partial G}{\partial n_v} = \frac{1}{A}(\Delta H_v - T\Delta S_v + RT \ln N_v)$$

and putting this equal to zero gives the equation for the mole fraction of vacancies

$$N_v = \exp\frac{\Delta S_v}{R} \exp\frac{-\Delta H_v}{RT} = \exp\frac{-\Delta G_v}{RT} \tag{4.8}$$

Eqn. 4.8 shows that at any temperature above absolute zero a crystal will have a finite concentration of vacancies, and that, as the temperature is increased, the mole fraction of vacancies also increases. The magnitude of the mole fraction depends on the size of the free energy of formation ΔG_v. Theoretical values of ΔS and ΔH have been calculated by Swalin (1961)*. For germanium and silicon, ΔS is

* Swalin, R. A.: *J. Phys. & Chem. Solids*, 1961, **18**, p. 290

about 4 eV K^{-1} and ΔH is about 2 eV. These values are substantially confirmed by the experimental work on the electrical conductivity of germanium containing quenched in vacancies.

A calculation similar to the one carried out above could be done for any defect, although the magnitudes of the mole fractions concerned could vary over many orders of magnitude depending on the free energy of formation. In particular, the calculation could be carried out for interstitials, using $\Delta G_i = \Delta H_i - T\Delta S_i$. The value of ΔG_i will depend critically on the crystal structure of the material concerned. A close-packed structure will have very little room for interstitial atoms, so ΔG_i would be expected to be large. More open lattices on the other hand, such as the diamond structure, will be able to accommodate interstitial atoms rather more easily, and ΔG_i will be smaller. This line of argument leads one to expect that interstitial atoms might have a role to play in diffusion in the many semiconductors that crystalise in the diamond structure, such as germanium, silicon and the III–V group. There is evidence to show that this is often true.

4.2 Defect complexes

Having established that point defects exist at equilibrium, we now consider associations of these defects, using the law of mass action. The simplest of these, since it does not require a second element, is the divacancy pair. This is described as an example, but it is by no means certain that divacancies occur to any great extent in germanium and silicon. Another simple interaction, that between a vacancy and an uncharged impurity atom will also be considered.

Vacancy pairs will exist whether or not there is any interaction between single vacancies. The laws of chance must occasionally cause vacancies, which are quite mobile, to appear on adjacent sites. However, there is reason to suppose that they can attract each other as well. There is an elastic-strain field associated with a vacancy, and often a divacancy should have a lower strain energy than a pair of single vacancies. The free energy would then be decreased on divacancy formation, and the concentration would be greater than random. Consider the reaction

$$2V \rightleftharpoons V_2$$

and assume the vacancies obey Henry's law. Applying the law of mass action

$$\frac{\gamma_{v_2} N_{v_2}}{(\gamma_v N_v)^2} = k \qquad (4.9)$$

Here N_{v_2} is the mole fraction of divacancies. No distinction is made here between N_v and N_{v1}, since $N_{v2} << N_{v1}$, so that N_{v1} approximately equals N_v. K is the equilibrium constant, given by

$$K = \exp \frac{-\Delta G}{RT}$$

and if the force between vacancies is one of attraction, ΔG must be negative. Eqn. 4.9 becomes

$$N_{v_2} = \frac{\gamma_v^{\,2}}{\gamma_{v_2}} N_v^{\,2} \exp\frac{-\Delta H}{RT} \exp\frac{\Delta S + \Delta S'}{R} \tag{4.10}$$

where ΔH is the difference in the enthalpy of formation between one mole of divacancies and two moles of single vacancies (in the interior of the crystal), i.e.

$$\Delta H = \Delta H_{v_2} - 2\Delta H_v$$

and, similarly,

$$\Delta S = \Delta S_{v_2} - 2\Delta S_v$$

$\Delta S'$ needs a little more explanation. It is an extra piece of configurational entropy over and above that already described for single vacancies. It allows for the fact that each lattice site in a crystal has Z nearest neighbour sites. A vacancy on a certain site has therefore Z chances to share in a divacancy. The number of ways that the crystal can accommodate a given number of divacancies is increased, and so, according to eqn. 4.1, the entropy is increased. The extra term is given by

$$\Delta S' = R \ln 0{\cdot}5Z$$

where $0{\cdot}5Z$ has been substituted for $\mathscr{W}$ rather than Z, because each orientation counted for a given vacancy is also an orientation for the other vacancy of the pair.

The point may be made a little more clear by the following. If we remove any interaction between vacancies tending to form divacancies, the ΔH and ΔS terms go to zero. Eqn. 4.10 becomes

$$N_{v_2} = N_v^{\,2} \exp\frac{\Delta S'}{R} = \frac{ZN_v^{\,2}}{2}$$

where an ideal solution has been assumed, so the γ terms are unity. Taking the example of the diamond lattice, which has co-ordination number $Z = 4$,

$$N_{v_2} = 2N_v^{\,2} \tag{4.12}$$

Since we have removed all interactions from the vacancies, eqn. 4.12 must represent the random coming together of vacancies in pairs. This can be calculated by a simpler method. The probability that a given site has a vacancy is given by the mole fraction N_v. The fraction of neighbouring sites containing a vacancy is also N_v, so the total fraction of sites occupied by a divacancy is $4N_v^{\,2}$ since there are four nearest neighbours. But each pair of such sites constitutes only a single divacancy, so the number must be divided by two to give $N_v^{\,2}$. This is the same result as eqn. 4.12.

The most important results of these considerations, however, are that the number of divacancies increases as the square of the vacancy concentration, and they become relatively more numerous at lower temperatures, since $\Delta G < 0$. The

0·5Z term, which will occur for other complexes also, is usually not considered, since it can be contained in the constant term in the mass-action formulation.

In a crystal containing impurity atoms, complexing can occur between these atoms and vacancies. The same criteria apply as for a divacancy: if the elastic energy of the lattice is lowered by the creation of the complexes, i.e. ΔG negative, they will form in numbers greater than random mixing would give. Impurity atoms are not in general the same size as the host atoms, and so they do not fit exactly into the crystal. They can often be accommodated more easily if they are situated next to an unoccupied site: for the same reason, impurities in crystals are attracted to dislocation lines. There are, therefore, two types of vacancies in the lattice; those that exist separately and those associated with an impurity. To a good approximation the two types exist independently of each other, the concentration of the first type being given by eqn. 4.8, and that of the second depending on the impurity concentration. This distinction would have been irrelevant for the divacancy complex, for which $N_{v1} >> N_{v2}$.

The complexing reaction between a foreign atom A and a vacancy can be written

$$V + A \rightleftharpoons VA$$

for which the mass-action equation is

$$\frac{\gamma_{vA} N_{vA}}{\gamma_v \gamma_A N_v N_A} = K \tag{4.13}$$

If all these defects are in sufficiently low concentration for Henry's law to be obeyed, the γ terms are constants, and

$$N_{vA} = K' N_v N_A \exp \frac{-\Delta G}{RT} \tag{4.14}$$

where K' includes the γ terms. As would be expected, the number of pairs is proportional to the product of the individual point-defect concentrations. If the force between the defects is attractive, as it must be for extra pairs to form, the importance of the complex increases as the temperature decreases. In a crystal in which vacancy–impurity defects occur, therefore, the total number of vacancies is dependent on the nature and concentration of the impurity atoms.

4.3 Electronic defects

Interactions between defects in crystals have so far been considered in terms of the relief of elastic strain causing attractive forces between defects. A more important factor in semiconductor crystals occurs as a consequence of many of the defects carrying charge, and therefore attracting and repelling each other. Associated with these charge phenomena is the requirement for charge neutrality. Electrons and holes count as charged defects in this context, as do charged atoms, vacancies and complexes.

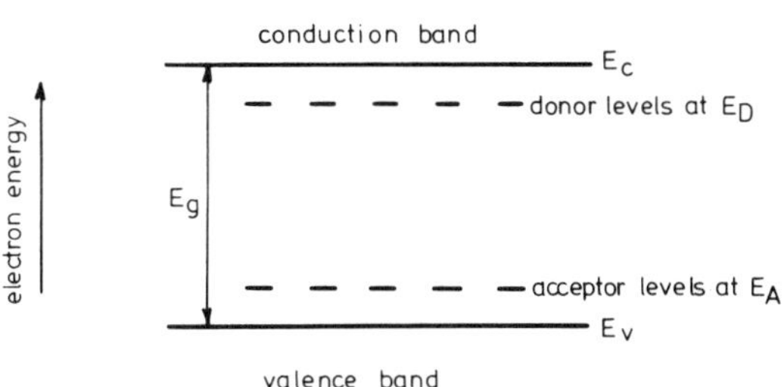

Fig. 4.1 Energy-band diagram for a semiconductor

The conventional way of representing these defects is the energy-band diagram shown in Fig. 4.1. The diagram shows the energies allowed to electrons in a crystal: it might refer to a semiconductor or an insulator, depending on the value of E_g. For a semiconductor, E_g is of the order of 1 eV. In a simple pure semiconductor, the allowed energies form into bands separated by forbidden gaps in which there are no allowed energy levels, or electron states. For the moment, ignore the donor and acceptor states (Fig. 4.1). The most important of these bands of allowed states are the valence-band and the conduction band. At 0 K all the states in the valence-band contain electrons, and we say that the valence-band is full. The conduction-band, however, is completely empty of electrons at 0 K. There are obviously no electrons in the energy range between the two bands, since, by definition, there are no electron states to be occupied.

When the valence-band is completely full and the conduction-band completely empty, the semiconductor cannot conduct electricity. Conduction-band electrons cannot conduct because there are none, and valence-band electrons cannot conduct because there are no spare accessible energy levels for an electron to move into. Thus if a valence-band electron increases its energy under the action of a battery, it must displace the electron that originally occupied the higher energy level. There is no level for the second electron to go to, except the one just left by the first electron. All that has happened, therefore, is that the two electrons have exchanged states. Since electrons are indistinguishable, this means that nothing has happened!*

When the temperature is raised above 0 K, a few electrons gain enough thermal energy to jump the gap from valence-band to conduction-band. This puts a relatively small number of electrons in conduction-band states, and leaves some empty states behind in the valence band. Electrons are then free to change their energies in both bands, and electrical conduction can take place. The lower band then contains a very large number of states occupied by electrons and a smaller number that are not. In considering the electrical properties of the semiconductor,

* The reader not familiar with these ideas will find them described in detail in books on solid-state physics. See, for example, Dekker, A. J., 'Solid state physics' (Macmillan, 1957)

it is desirable to work with the smaller number of unoccupied states, which behave, in practise, rather like positive electrons. These imaginary particles are called holes, and can be assigned an effective mass, which is usually of the order of the mass of an electron. In the semiconductor under discussion, in which there are no energy levels in the forbidden-gap, the numbers of electrons and holes must be equal. This type of semiconductor is called intrinsic.

When an electron returns to the valence-band, a hole is annihilated. In equilibrium at temperatures greater than absolute zero, the number of electrons going from valence-band to conduction-band is exactly balanced by the number going in the opposite direction. A quasichemical reaction can be written

$$eh \rightleftharpoons e + h \tag{4.15}$$

where e is a free electron in the conduction-band, h is a hole in the valence-band and eh is an electron-hole pair in the valence-band, or, to put it another way, is an occupied state in that band.

If impurity atoms are allowed into the crystal, they can provide new energy states within the forbidden-gap. These atoms can donate electrons to the conduction-band(donors), creating new electrons, or accept electrons from the valence-band(acceptors). The energies $E_C - E_D$, $E_A - E_V$ (Fig. 4.1) are the activation energies for donors and acceptors, respectively. If a crystal contains only donors, the effect is to create a situation in which $n > p$. In this event, the semiconductor is called n type. Similarly, a semiconductor containing only acceptors is called p type. In a crystal in which there are both donors and acceptors, the semiconductor is n or p type depending on which predominates. It is also possible for them to cancel each other out, forming a compensated crystal in which once again $n = p$.

The process introduces new charges into the semiconductor, since a donor that has donated its electron is obviously a positive ion. The neutrality condition becomes

$$n + N_A^- = p + N_D^+ \tag{4.16}$$

where n is the electron concentration, p is the number of holes and N_A^- and N_D^+ are the concentrations of ionised acceptors and donors, respectively. States can also be put into the forbidden-gap by native defects, such as vacancies and interstitials in some crystals. Then, even pure crystals have donor and acceptor levels.

A careful distinction must be drawn between a state on the energy diagram, and a state occupied by an electron. The conduction band, for instance, is always full of states, but at 0 K, it has no electrons at all. The probability that a state in one of the bands is occupied by an electron is given by the Fermi–Dirac distribution

$$F(E) = \frac{1}{1 + \exp\dfrac{E - E_F}{kT}} \tag{4.17}$$

where E is the energy of the state in question and E_F is called the Fermi energy or Fermi level. Its value is defined essentially by the requirement of eqn. 4.16. For semiconductors, the Fermi level is usually somewhere within the forbidden-gap. An n-type semiconductor has the level between the middle of the gap and the bottom of the conduction-band: a p-type sample has the level close to the top of the valence-band. The number of electrons within a certain energy range is therefore given by the product of the number of states within the range and the value of the Fermi–Dirac function at the appropriate energy. When calculating the number of electrons occupying states within the forbidden-gap, it is more correct to use a slightly modified form of the distribution. The number of electrons occupying N_D donor states of energy E_D within the gap becomes

$$n = \frac{N_D}{1 + \frac{1}{2}\exp\frac{E_D - E_F}{kT}}$$

where the factor ½ comes from the fact that a donor state can accept an electron of either clockwise or anticlockwise spin, but once occupied cannot be occupied further by an electron of opposite spin.

Electrons and holes must attain equilibrium in any semiconductor, so that eqn. 4.15 is valid whether the semiconductor is intrinsic, n type or p type. A mass-action relation can be written

$$\frac{\gamma_n \gamma_p np}{\gamma_{np}(np)} = K \tag{4.18}$$

The numbers of electrons and holes are very small compared to the number of full valance-band states (np). It is reasonable, therefore to assume Henry's law so that γ_n, γ_p are constants, and the activity of the full state, $a_{np}(np)$, is unity. Eqn. 4.18 simplifies to $np = BK$. The constant B includes the γ terms, and K depends only on temperature. This equation is very important in semiconductor work. It tells us that we can only increase the concentration of electrons, by doping with donors, at the expense of reducing the concentration of holes in the same ratio. When the material is intrinsic, then $n = p = n_i$, and the equation is often written

$$np = n_i^2 \tag{4.19}$$

The meaning of the constant K in eqn. 4.18 requires a little thought. So far we have taken constant temperature and pressure as standard conditions. This is reasonable, since it applies to the situation most commonly found in nature. K under these circumstances is given by eqn. 3.108, $\Delta G = -RT \ln K$. When dealing with a solution of electrons and holes in a solid it makes more sense to regard the solid as a solvent of constant volume. The Helmholz free energy F is then used rather than the Gibbs energy, and np is given by

$$np = B \exp\frac{-\Delta F}{RT} = B \exp\frac{\Delta S}{R} \exp\frac{-\Delta U}{RT} \tag{4.20}$$

here ΔF, ΔS, ΔU refer to the quantities required to create one mole of electron-hole pairs. Noting that the entropy term is a constant and dividing both ΔU and R by Avagadro's number to give $\Delta U'$ and k respectively, eqn. 4.20 can alternatively be written

$$np = B' \exp \frac{-\Delta U'}{kT} \tag{4.21}$$

and $\Delta U'$, the energy required to make a single electron-hole pair can be equated to E_g, the forbidden energy-gap in Fig. 4.1. In the same way, the ionisation of a donor

$$D \rightleftharpoons D^+ + e^-$$

has a mass-action equation giving rise to

$$nN_D^- = B'' \exp \frac{-\Delta U''}{kT}$$

and the energy $\Delta U''$ is the ionisation energy for the donor, equal to $(E_c - E_D)$.

The difference between using ΔF and ΔG is not as great as it might appear. From the definitions

$$\Delta G = \Delta H - \Delta(TS)$$

$$\Delta F = \Delta U - \Delta(TS)$$

Now enthalpy is given by

$$H = U + PV$$

so any change in enthalpy due to ionisation is

$$\Delta H = \Delta U + \Delta(PV)$$

For a solid, $\Delta(PV)$ is a very small quantity, so that $\Delta H \simeq \Delta U$ and $\Delta G \simeq \Delta F$.

4.3.1 Statistical approach

An equation of the form of eqn. 4.21 is frequently used in semiconductor work, but it is usually derived using Fermi–Dirac statistics rather than the law of mass action. It is instructive to review the statistical calculation because certain explicit assumptions have to be made to obtain an expression of the desired form. Assumptions were also made in deriving eqn. 4.21. We should be able to say, therefore, that the two sets of assumptions are different statements of the same ideas.

The density of electron states in the conduction-band in the energy range E to $E + dE$ is

$$S(E)\, dE = \frac{4\pi}{h^3} (2m_e)^{3/2} (E - E_C)^{1/2}$$

where m_e is the effective mass of an electron and h is Planck's constant. The number of electrons in this range is

$$N(E)dE = F(E) \cdot S(E) \cdot dE$$

where $F(E)$ is given by eqn. 4.17. To find the number of electrons in the conduction-band, it is necessary to integrate this expression over the energy range covered by the conduction-band. The energy of the bottom of the band is E_C and the function $F(E)$ decreases so sharply with increasing energy that it makes little difference what value is assigned to the top of the band. We will take infinity. Hence

$$n = \frac{4\pi}{h^3}(2m_e)^{3/2}\int_{E_C}^{\infty}(E - E_C)^{1/2} \cdot \frac{1}{1 + \exp\dfrac{E - E_F}{kT}} \cdot dE \tag{4.22}$$

This function cannot be integrated explicitly. However, often E_F is several kT less than the lowest energy considered, i.e. $E_C - E_F > 4kT$, say. The exponential term in the denominator then predominates, and the Fermi–Dirac function can be approximated to

$$F(E) \simeq \exp\frac{E_F - E}{kT}$$

The approximation being made is that of assuming that the Fermi–Dirac distribution can be replaced by the 'classical' Boltzman distribution. Making the substitution gives

$$n = \frac{4\pi}{h^3}(2m_e)^{3/2}\exp\frac{E_F}{kT}\int_{E_C}^{\infty}(E - E_C)^{1/2}\exp\frac{-E}{kT}\,dE$$

By changing the origin of the energy co-ordinates to E_C and making the substitution $x = E/kT$, we obtain

$$n = \frac{4\pi}{h^3}(2m_e kT)^{3/2}\exp\frac{E_F - E_C}{kT}\int_0^{\infty}x^{1/2}e^{-x}\,dx$$

The integral is standard and comes to $\frac{1}{2}\pi^{1/2}$, hence

$$n = 2\left(\frac{2\pi m_e kT}{h^2}\right)^{3/2} \cdot \exp\left(\frac{E_F - E_C}{kT}\right) \tag{4.23}$$

A similar procedure for holes gives

$$p = 2\left(\frac{2\pi m_h kT}{h^2}\right)^{3/2} \cdot \exp\left(\frac{E_V - E_F}{kT}\right) \tag{4.24}$$

Putting for simplicity $m_e = m_h = m$

$$np = 4\left(\frac{2\pi mkT}{h^2}\right)^3 \exp\frac{-E_g}{kT} \tag{4.25}$$

where $E_C - E_V = E_g$. Eqn. 4.25 is of the same form as eqn. 4.21, as expected.

In deriving eqn. 4.21 the assumption was made that Henry's law is obeyed, so that the activity coefficient terms can be treated as constants. The statistical treatment shows that such an assumption is justified only in those circumstances in which the electrons and holes follow classical statistics. This is quite a stringent condition: for heavily doped semiconductors it is usually necessary to employ Fermi–Dirac statistics. Some care must therefore be used when employing solution theory to quasichemical reactions involving electrons and holes. If the law of mass action is to be employed for concentrations in which Henry's law is not obeyed, a variable activity coefficient must be used.

4.3.2 Activity coefficient for electrons

We have shown that eqn. 4.21 is correct providing Henry's law can be assumed, and this in turn relies on the electrons and holes obeying Boltzman statistics. Unfortunately the Boltzman approximation can break down at quite low concentrations of electrons and holes, giving rise to the difficulty mentioned above. If we call the actual concentration of electrons n_e, and the concentration predicted by Boltzman statistics n_B, then it is n_B that is required in the mass-action formulation. The concentration n_e must therefore be multiplied by some function γ to transform it to n_B, i.e.

$$n_B = \gamma n_e \tag{4.26}$$

The term γn_e is then used in the mass-action equation. It can be seen that γ is simply the electron activity coefficient. It will be a variable at electron concentrations which are too great for Henry's law to be obeyed. Eqn. 4.26 allows us to derive an expression for γ. The quantity n_B is given by eqn. 4.23. There is no absolute scale of energy, so we can choose $E_C = 0$, since it makes the calculation easier. Eqn. 4.23 can be written

$$n_B = \lambda \exp \eta \tag{4.27}$$

λ represents the pre-exponential term in eqn. 4.23 and $\eta = E_F/kT$.

The exact expression for the number of electrons is given by eqn. 4.22. Again choose $E_c = 0$, and the expression can be further simplified by the substitution $x = E/kT$. The equation becomes

$$n_e = \frac{2}{\pi^{1/2}} \cdot \lambda \int_0^\infty \frac{x^{1/2}}{1 + \exp(x - \eta)}\, dx \tag{4.28}$$

It has already been noted that this integral cannot be solved explicitly, but this can be done numerically. Call the solution $F(\eta)$. Substituting eqns. 4.28 and 4.27 into eqn. 4.26 then gives an expression for γ

$$\gamma = \frac{\pi^{1/2}}{2} \cdot \frac{\exp \eta}{F(\eta)} \tag{4.29}$$

This equation has been evaluated by Rosenberg (1960)* using numerical solutions for $F(\eta)$. Fig. 4.2 shows his results, plotting γ against both η (normalised Fermi energy) and against n_e/λ. The two plots amount to the same thing, since η and n_e/λ are uniquely related by eqn. 4.28. The latter axis is the more useful, however, since it indicates at which electron concentrations Henry's law does not apply. The function λ is given by

$$\lambda = 2\left(\frac{2\pi m_e kT}{h}\right)^{3/2}$$

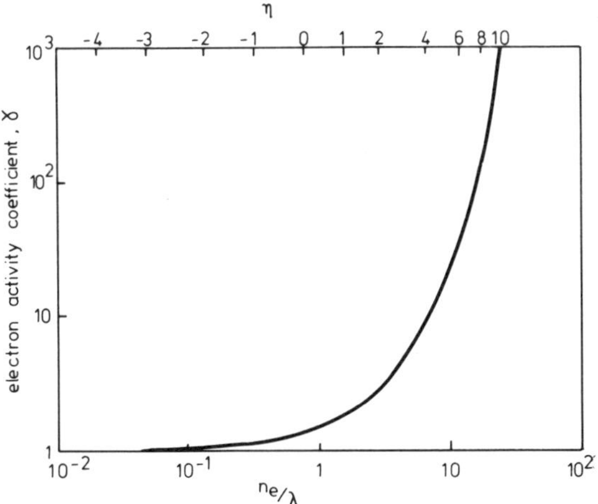

Fig. 4.2 Activity coefficient for electrons, as a function of electron concentration

Reproduced from Rosenberg, A. J.: *J. Chem. Phys.*, **1960, 33**, p. 665

Its value depends on the electron effective mass and will be different for different semiconductors. Taking m_e as the normal free-electron mass, λ has a value $4{\cdot}8 \times 10^{21}\ T^{3/2}\ m^{-3}$, which, at 300K comes to $2{\cdot}5 \times 10^{25}\ m^{-3}$. Some semiconductors have a value of m_e as low as one tenth of the normal electron mass, m, reducing λ by a factor $(m/m_e)^{3/2}$.

* Rosenberg, A. J.: *J. Chem. Phys.*, 1960, **33**, p. 665

Fig. 4.2 shows that γ exceeds unity by a considerable amount for concentrations greater than $n_e/\lambda \simeq 1$. After this point the rise is almost exponential. Reference to the other axis, on which η is plotted, shows that the sharp rise in γ roughly corresponds to $\eta > 0$. Noting that the bottom of the conduction-band was taken at the zero of energy, $\eta = 0$ corresponds to the point at which the Fermi level is at the bottom of the band. A semiconductor in which the Fermi level is actually within either the conduction-band or the valence-band is called degenerate, and it can be seen from the Figure that deviation from Henry's law roughly corresponds to the onset of degeneracy in the semiconductor. (It is also in this region that the classical approximation to Fermi–Dirac statistics becomes invalid.) Note that since η and λ are functions of temperature, the degeneracy point also depends on the temperature. This means that for diffusion temperatures, degeneracy occurs at concentrations rather higher than those quoted above, since these temperatures are typically much greater than 300 K.

An exactly analogous argument could be carried through for holes in the valence band, defining an activity coefficient for holes. The starting point would be eqn. 4.24 rather than eqn. 4.23, but the same result would be obtained and Fig. 4.2 applies equally well. The effective mass for holes must now be used in calculating λ.

4.3.3 Chemical potential for electrons

In considering both the usual physical and chemical ways of looking at materials, it is interesting to see the way in which two treatments which look formally very different often express the same idea. A good example of this is to be found in the apparently quite different concepts of chemical potential of an electron and Fermi energy; which turn out to be exactly the same. This is demonstrated by deriving expressions for the change in entropy of a crystal when a single electron is added.

Suppose the electron is brought from infinity to occupy an energy level E_j in the crystal. Let the number of states at this energy be Z_j. Suppose there are initially n_j electrons at this energy. Their entropy is given by eqn. 4.1 putting $\mathscr{W}$ equal to the number of ways in which n_j electrons can occupy Z_j states,

$$S = k \ln \frac{Z_j!}{n_j!(Z_j - n_j)!}$$

Using Stirling's approximation, this comes to

$$S = Z_j \ln Z_j - n_j \ln n_j - (Z_j - n_j) \ln (Z_j - n_j)$$

and

$$\frac{dS}{dn_j} = k \ln \left(\frac{Z_j - n_j}{n_j}\right)$$

Now Z_j and n_j are related by the Fermi function

$$n_j = F(E_j)Z_j$$

giving

$$\frac{dS}{dn_j} = k \ln \left[\frac{1 - F(E_j)}{F(E_j)} \right] \tag{4.30}$$

which gives the entropy change per added electron. Substituting in eqn. 4.30 for the Fermi function gives the extra entropy due to the addition of one electron as

$$dS = -\frac{E_F}{T} + \frac{E_j}{T} \tag{4.31}$$

We now require a second expression for this entropy change. Any thermodynamic property can be expressed as a function of the number of particles plus two other thermodynamic properties. Let us express entropy in terms of the total number of particles N, volume, and energy, and add the restriction of constant volume, so there is no $\partial S/\partial V$ term. The entropy change due to the addition of dN particles at energy U may be written

$$dS = \left(\frac{\partial S}{\partial N}\right)_{U,V} dN + \left(\frac{\partial S}{\partial U}\right)_{N,V} dU$$

If a single electron is brought up from zero energy to E_j, the change in the number of particles dN is one, and the change in total energy dU is E_j. Thus

$$dS = \left(\frac{\partial S}{\partial N}\right)_{U,V} + \left(\frac{\partial S}{\partial U}\right)_{N,V} \cdot E_j \tag{4.32}$$

Eqn. 4.32 must now be compared to eqn. 4.31. The comparison gives the following relationships:

$$\left(\frac{\partial S}{\partial N}\right)_{U,V} = -\frac{E_F}{T} \tag{4.33}$$

$$\left(\frac{\partial S}{\partial U}\right)_{N,V} = \frac{1}{T} \tag{4.34}$$

The first of these can be related to chemical potential. In the previous chapter, one of the definitions of μ was

$$\mu = \left(\frac{\partial F}{\partial N}\right)_{V,T}$$

Since

$$F = U - TS$$

$$\left(\frac{\partial F}{\partial N}\right)_{V,T} = \frac{\partial U}{\partial N} - T\frac{\partial S}{\partial N}$$

making $(\partial S/\partial N)$ the subject of the equation and adding the further restriction of constant energy U

$$\left(\frac{\partial S}{\partial N}\right)_{U,V} = -\frac{1}{T}\left(\frac{\partial F}{\partial N}\right)_{V,T} = -\frac{\mu}{T} \tag{4.35}$$

Comparison of eqns. 4.33 and 4.35 shows that

$$\mu_e = E_F \tag{4.36}$$

It is necessary at this stage to point out a possible ambiguity. The chemical potential quoted in eqn. 4.36 is the Gibb's free energy of a single electron. Normally in physical chemistry, the unit amount of material dealt with is the mole. This is a reasonably large amount, bearing some relation to the amounts of material used in practise. The single atom is rarely used as the basic unit. In semiconductor physics, the single electron and single hole are, however, taken as the unit. The energies quoted on a diagram such as Fig. 4.1 refer to a single electron: E_F is the energy of an electron at the Fermi level. When dealing with defects in crystals, some of which will be atomic and some electronic, it is usually more convenient to talk in terms of single defects. There seems to be no convention about this, however. The simplest way to decide whether a given argument refers to moles or single defects is to note whether Boltzman's constant, k, or the gas constant, R, is used. The former is usually found in atomic derivations and the latter in molar ones.

Finally, consider two samples of a semiconductor containing different dopings of donors and acceptors, so that the electron and hole concentrations in the two samples are different. The activities and chemical potentials of the electrons in the two samples will also be different. If the samples are now brought into intimate contact, then a flow of electrons between the two will take place to make the activity and chemical potential of electrons in sample 1 equal to those in sample 2. At equilibrium we have $\mu_1 = \mu_2$, so that $E_{F1} = E_{F2}$. Thus we arrive at the familiar statement that the Fermi level is continuous across a semiconductor junction at equilibrium, giving another example of the different ways in which the languages of physics and chemistry express the same idea.

4.4 Interactions between charged defects

Because of the presence of electrons and holes and the fact that many atomic defects carry a charge, interactions between charged defects are especially important. Before going any further into the subject of defect complexes, however, it is helpful to modify the nomenclature somewhat. It proves to be more convenient to describe concentrations of crystal defects in terms of defects per unit volume, rather than mole fractions. The two sets of units are obviously related by a constant ratio; the mole fraction of vacancies in silicon, for instance, is equal to the number of silicon vacancies per unit volume divided by the number of lattice sites in that volume. The law of mass action is still applicable, therefore, with the equilibrium constant multiplied by the appropriate factor. The concentration of a species can be simply represented by putting a set of square brackets about the symbol for the species. Thus if V indicates a vacancy, the concentration of vacancies per unit volume is written $[V]$; similarly, the concentration of ionised donors might be written $[D^+]$.

To return to the interactions: they can be divided into two main types. In the first, complexing takes place between defects of opposite sign by the simple mechanism of coulombic attraction. This sort of interaction will be considered later in the chapter. The second type, which is rather more subtle, is the more important in semiconductors and will be described in some detail. It relies on the fact that the numbers of electrons and holes, and the degree of ionisation of acceptors and donors all depend on the position of the Fermi level. The Fermi level itself, however, depends in turn on the concentrations of donors and acceptors. The concentration of donors therefore influences the degree of ionisation of acceptors and also, as we shall see, their solubility. The interaction takes place not, therefore, by way of physical proximity but, to use chemist's language, through the law of mass action.

As an example of this type of intereaction, consider the vacancy concentration $[V]$ in an elemental semiconductor. It is usually assumed that such vacancies act as acceptors and for germanium there is clear evidence that this is true. At any temperature the crystal has a definite number of uncharged vacancies $[V^0]$ say, given by eqn. 4.8. However, since a vacancy can act as an acceptor, there will also be a concentration of charged vacancies $[V^-]$ and the total concentration will be given by $[V] = [V^0] + [V^-]$. If the energy level of the acceptor is known, then $[V^-]$ and $[V]$ can be related via the Fermi–Dirac function. However, it is simpler to tread the mass-action path and describe the ionisation process, thus:

$$V^0 \rightleftharpoons V^- + h$$

where V^0 is an uncharged vacancy and V^- is charged. Hence

$$p[V^-] = K[V^0]$$

where Henry's law is assumed, and any (constant) activity coefficient terms are swallowed up in K. The hole concentration can be replaced by n_i^2/n, and since n_i is a constant at constant temperature, we can write

$$[V^-] = K_1[V^0]n \tag{4.37}$$

That is, at a fixed temperature (at which $[V^0]$ is constant) the number of charged vacancies is proportional to the number of free electrons. Therefore an addition of donors will increase the charged-vacancy density, leading to the interesting effect, from the electronic point of view, of self compensation, i.e. adding more donors to the crystal causes it to produce more acceptors. Contrariwise, the addition of acceptors decreases the charged vacancy concentration. In terms of the total numbers of atoms the effect is small, since the number of vacancies is always small compared to the number of atoms. Some phenomena depend critically on the number of vacancies, however, and the interaction effect will be important for them. The most obvious case of a quantity which will be affected is the self diffusion coefficient of the element. Assuming the diffusion mechanism to be one of simple vacancy exchange, the diffusion coefficient will be proportional to the vacancy concentration. Now if the vacancy acceptor level is close to the top of the valence band most of the vacancies will be ionised and $[V^0] \ll [V^-]$. This gives $D \propto n$, or

$$\frac{D}{D_i} = \frac{n}{n_i} \tag{4.38}$$

where D_i is the self diffusion coefficient measured in intrinsic material. Eqn. 4.38 has been checked by Valenta and Ramasastry (1957)*, who carried out self diffusion experiments in intrinsic germanium and also in heavily doped n-type and p-type material. Their results are shown in Fig. 4.3. It is clear from the Figure that the diffusion coefficient in n-type material is greater than that in intrinsic or p-type material. As the temperature is increased, the difference becomes smaller. This is a consequence of the fact that at higher temperatures the Fermi level moves towards the centre of the forbidden-gap. In intrinsic material the Fermi level is always at the centre of the gap: as the temperature is raised, both n and p type therefore approach the 'intrinsic' state, and $n \rightarrow n_i$. Thus the vacancy concentrations converge and so do the diffusion coefficients.

We now consider the way in which the donor doping of a crystal can influence the solubility of an acceptor impurity and vice-versa. Suppose that we have a semiconductor doped with a donor D, and that it is in equilibrium with some external phase containing D. Let D in this external phase be called D_{ext}. A reaction can be written

$$D_{ext} \rightleftharpoons D \rightleftharpoons D^+ + e \tag{4.39}$$

in which the latter part of the equation refers to donor ionisation. The free electron produced can react with the holes present

$$e + h \rightleftharpoons eh \tag{4.40}$$

In terms of mass-action, an acceptor should increase the donor solubility, since it produces holes via the reaction

$$A \rightleftharpoons A^- + h \tag{4.41}$$

The holes can react with electrons by way of eqn. 4.40, tending to exhaust the electron concentration. This has the effect of driving eqn. 4.39 to the right, causing

* Valenta, M. W., and Ramasastry, C.: *Phys. Rev.*, 1957, **106**, p. 73

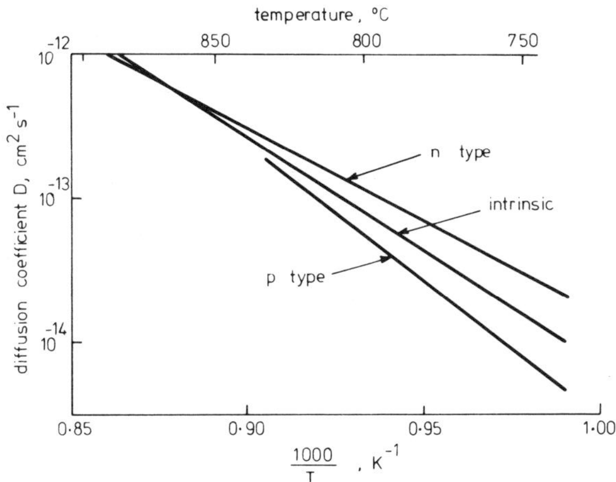

Fig. 4.3 Self-diffusion coefficient of germanium as a function of temperature for three specimens. The specimens were:

a *n* type, doped to a level 6×10^{18} cm^{-3}
b intrinsic
c *p* type, 1×10^{20} cm^{-3}

Reproduced from Valenta, M. W., and Ramasastry, C.: *Phys. Rev.,* 1957, **106**, p. 73

more donors to become dissolved in an attempt to replace the electrons. In the same way, the presence of one donor should decrease the solubility of another. This can be seen clearly if we write down the mass-action equations.

Let $[D]$ be the total concentration of donors, and $[D^+]$ be the concentration of ionised donors, with $[A]$, $[A^-]$ as the corresponding acceptor concentrations. If the activity of the donor in the external phase is a, the first part of eqn. 4.39 can be represented

$$\frac{[D] - [D^+]}{a} = K_1 \tag{4.42}$$

The second part has, similarly,

$$\frac{[D^+]n}{[D] - [D^+]} = K_2 \tag{4.43}$$

Eqn. 4.40 gives

$$np = n_i^2 \tag{4.44}$$

where n_i is a function only of temperature: it is shown in Fig. 4.4 for germanium and silicon. Substitution of eqn. 4.43 in 4.42 gives

$$[D^+]n = aK_1K_2 = K_3 \tag{4.45}$$

And for electrical neutrality of the crystal

$$[D^+] + p = [A^-] + n \tag{4.46}$$

The donors and acceptors of interest are all likely to be 'shallow', i.e. the donor levels near to the bottom of the conduction-band and the acceptor levels close to the top of the valence-band. Since we are assuming a 'classical' system in which Boltzman statistics are valid, the Fermi level must be further than several kT away from both bands. It follows, therefore, that to a good approximation, all donors and acceptors are ionised, so we can put

$$[D] = [D^+]$$

$$[A] = [A^-]$$

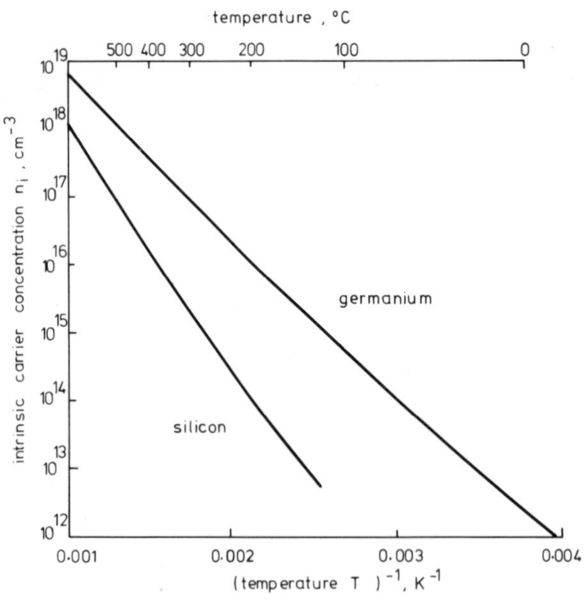

Fig. 4.4 Intrinsic carrier concentrations of silicon and germanium as functions of temperature

Reproduced from Reiss, H., Fuller, C. S., and Morin, F. J.: *Bell Syst. Tech. J.,* 1956, **35**, p. 533

Eqns. 4.44, 4.45 and 4.46 can now be used to derive an expression for the donor solubility $[D]$. This involves finding an expression for K_3.

Let the value of $[D^+]$ when there are no acceptors present in the crystal be $[D_0{}^+]$. Solving the equations for this special case, we can first add eqns. 4.44 and 4.45.

$$K_3 + n_i^2 = [D_0{}^+]n + np = n([D_0{}^+] + p) \tag{4.47}$$

now, when there are no acceptors, eqn. 4.46 becomes

$$[D_0{}^+] + p = n$$

putting this in eqn. 4.47 gives

$$K_3 + n_i^2 = n^2 = \frac{K_3{}^2}{[D_0{}^+]^2}$$

This equation gives K_3 in terms of parameters which are, in principle, known, namely n_i and $[D_0{}^+]$, i.e.

$$K_3 = \frac{[D_0{}^+]^2}{2} + \left\{\frac{[D_0{}^+]^4}{4} + n_i^2[D_0{}^+]^2\right\}^{1/2} \tag{4.48}$$

We now have three equations; eqns. 4.45, 4.46 and 4.48, from which the two variables n and K_3 can be eliminated to give $[D]$ (equal to $[D^+]$) in terms of $[A]$, which is the same as $[A^-]$. The algebra is somewhat tedious, but does eventually give a useful result

$$D = [D^+] = \frac{[A^-]}{y} + \left\{\left(\frac{[A^-]}{y}\right)^2 + [D_0{}^+]^2\right\}^{1/2} \tag{4.49}$$

where

$$y = 1 + \left\{1 + \left(\frac{2n_i}{[D_0{}^+]}\right)^2\right\}^{1/2}$$

If the acceptor concentration is to influence the donor solubility to any marked extent, it is necessary for the first term in the bracket of eqn. 4.49 to predominate since, if the reverse is true and $[D_0{}^+]^2 \gg ([A^-]/y)^2$, the bracket reduces to $[D_0{}^+]$ and the expression to $[D^+] = [D_0{}^+]$. The effect is observed, therefore if $[A^-] > y[D_0{}^+]$.

This can be written

$$[A^-] > [D_0{}^+] + ([D_0{}^+]^2 + 4n_i^2)^{1/2} \tag{4.50}$$

On the right-hand side of eqn. 4.50 either the term $[D_0{}^+]^2$ will predominate or the term n_i^2. In the first case, the inequality comes to $[A^-] > 2[D_0{}^+]$; in the second to $[A^-] > 2n_i$. The condition for a significant solubility effect can therefore be written

$$[A^-] > \left.\begin{matrix} 2n_i \\ 2[D_0{}^+] \end{matrix}\right\} \text{ whichever is the greater} \tag{4.51}$$

Now consider the situation when eqn. 4.51 is obeyed. The second term in the bracket of eqn. 4.49 can be neglected in comparison with the first and the equation becomes

$$[D^+] = \frac{2[A^-]}{y} \tag{4.52}$$

Since y is constant at constant temperature, this means that the donor solubility is proportional to the acceptor concentration at constant temperature. The variation of $[D^+]$ with temperature in eqn. 4.52 comes from the term $(2n_i/[D_0{}^+])^2$ in y. Since both n_i and $[D_0{}^+]$ increase with temperature, the variation is likely to be small. In the region for which eqn. 4.52 describes the donor solubility, the temperature variation is therefore less than that in the region in which the condition given by eqn. 4.51 is not obeyed.

Eqn. 4.49 has been checked by Reiss, Fuller and Morin (1956)*. A set of their results is shown in Fig. 4.5, in which the solubility of the donor lithium is plotted as a function of concentration of the acceptor boron. The points are experimental and the lines are plotted from theory. The external phase used was tin containing 0·18 percent lithium by weight. All of the features mentioned above are apparent in the Figure. The increase in donor solubility due to the presence of an acceptor is really quite striking: the Figure shows an increase by a factor of 1000 in one case. At low values of $[A^-]$ the donor doping is insensitive to acceptor concentration, but at high $[A^-]$, the relationship is linear. Note also that the temperature dependence is much less marked in the linear region.

4.4.1 Physical association in ion pairs

Donor and acceptor ions must obviously attract each other by columbic force and there must therefore be some tendency for them to occupy adjacent lattice sites to form ion pairs. Suppose the number of pairs formed is $\mathscr{P}$. The total number of donors, $[D]$ is no longer equal to the number of donor ions $[D^+]$, since $\mathscr{P}$ donors are involved in pairing, i.e.

$$[D] = [D^+] + \mathscr{P}$$

and

$$[A] = [A^-] + \mathscr{P}$$

* Reiss, H., Fuller, C. S., and Morin, F. J.: *Bell System Tech. J.*, 1956, **35**, p. 535

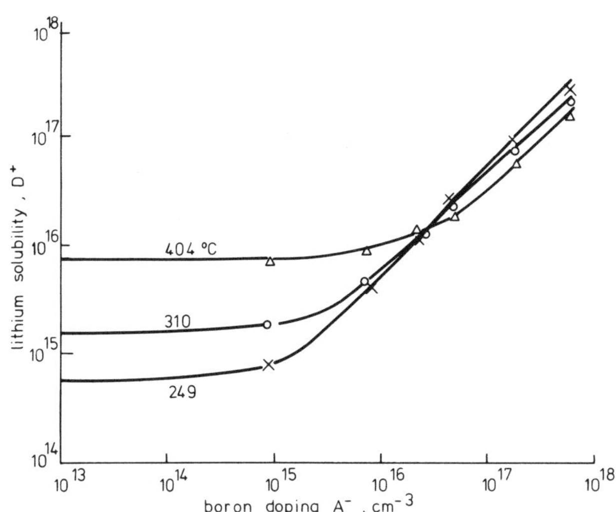

Fig. 4.5 Solubility of lithium in silicon as a function of boron doping, taken for three temperatures

Reproduced from Reiss, H., Fuller, C. S., and Morin, F. J.: *Bell Syst. Tech. J.,* 1956, **35**, p. 533

The mass-action equation for pairing is

$$\frac{\mathscr{P}}{[A^-][D^+]} = \Omega$$

where Ω is the mass-action constant. In addition, the equations for electron equilibrium and charge neutrality, eqns. 4.44, 4.45 and 4.46 must still hold

$$np = n_i^2 \tag{4.44}$$

$$n[D^+] = K \tag{4.45}$$

$$[D^+] + p = [A^-] + n \tag{4.46}$$

Again, the problem is to solve these six equations to give the donor solubility $[D]$ as a function of $[A]$. Note that if the pairing constant Ω is put equal to zero the problem reduces to the previous one, so the previous example might be considered to be a special case of this one.

The problem has to be solved numerically and the matter will not be further pursued here, except to give the results of some of the calculations of Reiss *et al.** They took as their example germanium doped with the substitutional acceptor gallium, and calculated the solubilities of the interstitial acceptor lithium. They also found the number of lithium-gallium ion pairs. A set of results for 300 K is shown in the Table 4.1. The first column gives the gallium concentration and the second the corresponding lithium solubility (including both single and paired atoms). The third column gives the answer for $[D]$, which is obtained by putting $\Omega = 0$, i.e. no pairing at all. This is the answer which would have been obtained in the previous section. The final column gives the number of pairs. At low gallium doping the effect of pairing is small, but at $[A] = 10^{18}$ cm^{-3}, about 98 percent of the lithium ions are involved in pairs. It is interesting that the numbers in the second and third columns are never very different, i.e. the effect of pairing is not to change the total number of donor atoms very much. This explains why the theoretical curve of Fig. 4.5 agreed so well with the experiments, even though pairing phenomena were not taken into account in the theory leading up to eqn. 4.49.

It is clear that at 300 K pairing can be important when there are high concentrations both of donors and acceptors. The way this effect varies with temperature is shown in Fig. 4.6; a graph plotted assuming equal numbers of donors and acceptors. Only at low temperatures and high concentrations are the majority of ions involved in pairs. Since a typical diffusion temperature for germanium would be 800°C, it would seem that ion pairs are not a major factor in diffusion, at least in this system. The high degree of association predicted for low temperatures must be looked at fairly closely. The data of Fig. 4.6 comes from equilibrium calculations, and it has been observed before that at low temperatures equilibrium can take an extremely long time to come about. The extra pairs that should form at low temperature can only come about by one ion diffusing through the crystal to a

Table 4.1 Concentration of pairs in germanium doped with both gallium and lithium. Calculated using the data:

$T = 300$ K

$n_i = 2{\cdot}8 \times 10^{13}$ cm^{-3}

$[D_0{}^+] = 7 \times 10^{13}$ cm^{-3}

$\Omega = 1{\cdot}6 \times 10^{15}$ cm^{-3}

(from Reiss, H., Fuller, C. S., and Morin, F. J.: *Bell Syst. Tech. J.*, 1956, **35**, 533)

[A], cm^{-3}	[D], cm^{-3}	[D]*, cm^{-3}	$\mathscr{P}$ = [A] − [A$^-$], cm^{-3}
10^{14}	$1{\cdot}25 \times 10^{14}$	$1{\cdot}25 \times 10^{14}$	$0{\cdot}15 \times 10^{14}$
10^{15}	$0{\cdot}94 \times 10^{15}$	$0{\cdot}875 \times 10^{15}$	$0{\cdot}44 \times 10^{15}$
10^{16}	$0{\cdot}985 \times 10^{16}$	$0{\cdot}875 \times 10^{16}$	$0{\cdot}77 \times 10^{16}$
10^{17}	$0{\cdot}990 \times 10^{17}$	$0{\cdot}875 \times 10^{17}$	$0{\cdot}92 \times 10^{17}$
10^{18}	$0{\cdot}995 \times 10^{18}$	$0{\cdot}875 \times 10^{18}$	$0{\cdot}97 \times 10^{18}$

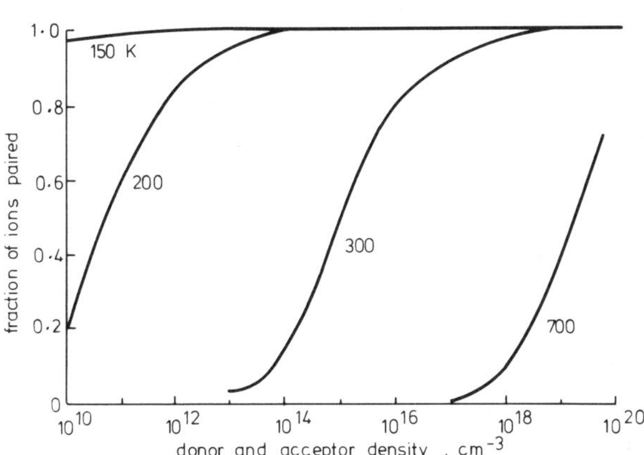

Fig. 4.6 Fraction of ions paired in germanium, assuming equal densities of positive and negative ions

Reproduced from Reiss, H., Fuller, C. S., and Morin, F. J.: *Bell Syst. Tech. J.,* 1956, **35,** p. 533

position adjacent to another ion, and diffusion is usually a very slow process at low temperatures. In the particular example chosen, the donor is lithium, which exists interstitially in germanium and is quite mobile down to low temperatures. For a donor and acceptor pair that both exist substitutionally in the lattice, it does not seem possible that the high concentration of pairs predicted by the theory for low temperatures could occur within a reasonable time scale.

4.5 Defects in compound semiconductors

The principles outlined above for elemental semiconductors apply just as well to a compound *MX*, say. The existence of two different types of atoms in the lattice gives a little more scope to the defects, however, and the following extra ones might be observed:

(*a*) Two types of vacancies, namely those for *M* and *X* lattice sites respectively. Similarly there are both *M* interstitials and *X* interstitials

(*b*) Atoms occupying sites assigned to atoms of a different type, i.e. *M* atoms on *X* sites and vice-versa

(*c*) Foreign atoms on either *M* or *X* sites.

In addition, a new set of defect complexes is possible such as an *M* vacancy, *X* vacancy pair, or various types of impurity-vacancy pairs. The further possibility exists of the total number of M atoms not corresponding exactly to the total of X atoms, i.e. nonstoichiometry. These variations occur in nature and are sometimes large: the compound FeO, for instance, can exist with as much as 54 atomic percent oxygen and is unstable, in fact, when it is composed of exactly 50 percent of each type of atom. Most compounds have a very small range of nonstoichiometry, but it can be significant when considered against the small numbers of defects we are dealing with anyway. The particular composition that a solid has, within its allowed range, turns out to be determined by the vapour pressures of its constituents. This point will be enlarged upon below.

Let us initially consider the defect equilibrium in a compound MX in which the defects are not charged. Suppose the species of X in the ambient vapour is X_2 and that the other element appears simply as M. If the only crystal defects of significance are M and X vacancies, the reaction may be written

$$(1-\delta)MX + \delta M(\text{vapour}) \rightleftharpoons MX + \delta V_X \tag{4.53}$$

where V_X is a vacancy on an X site. The equation simply states that the addition of a small mole fraction of extra M atoms, δ, can be considered as increasing the size of the crystal and adding mole fraction δ vacancies. Now since MX is a pure substance in its standard state, its activity is unity. The mass-action equation for eqn. 4.53 is therefore quite simply

$$\frac{[V_X]}{P_M} = K_1 \tag{4.54}$$

Where once again it proves to be more convenient to write down the relation in terms of concentrations rather than mole fractions. The reader will remember that this is quite permissible and merely has the effect of changing the K term by a constant factor. For vacancies on the other site the corresponding equation is

$$2(1-\delta)MX + \delta X_2(\text{vapour}) \rightleftharpoons 2MX + 2\delta V_M$$

giving rise to

$$\frac{[V_M]^2}{P_{X_2}} = K_2 \tag{4.55}$$

Finally, we can write

$$2M(\text{vapour}) + X_2(\text{vapour}) \rightleftharpoons 2MX$$

and

$$P_M{}^2 \cdot P_{X_2} = K_3 \tag{4.56}$$

Eqn. 4.56 expresses the idea that the vapour pressures above a compound are not independent, so that if the temperature and one of the pressures are chosen, the other pressure is fixed. This is a general statement of the relationship found for

GaAs in Chapter 3. Eqns. 4.54 and 4.55 show that the vacancy concentrations (and therefore the stoichiometry of the crystal) depend on the external vapour pressures. Since these pressures are not independent, neither are the vacancy concentrations. Using eqns. 4.54 and 4.55 to eliminate the pressures from eqn. 4.56 yields

$$[V_M]\,[V_X] = K_4 \tag{4.57}$$

The constant K_4 depends only on temperature, so eqn. 4.57 is true for the crystal at all points within the stoichiometry allowed by its phase diagram. If some of the vacancies are charged, eqn. 4.57 remains as the relationship between the uncharged species.

If the major defects are interstitial atoms, M_i and X_i say, then a similar argument can be followed through. The removal of M atoms from the lattice to interstitial sites can be described by the reaction

$$MX \rightleftharpoons M_{1-\delta}X + \delta V_M + \delta M_i$$

Now since by postulate $\delta \ll 1$, the activities both of MX and of $M_{1-\delta}X$ are unity. The equation therefore gives

$$[V_M]\,[M_i] = K_5 \tag{4.58}$$

and there is also a similar equation

$$[V_X]\,[X_i] = K_6 \tag{4.59}$$

Because the pure compound does not give rise to a term in the mass-action equation, it is possible to achieve the same result by writing down the original quasichemical reaction using only the defect species, i.e. as

$$M_M \rightleftharpoons V_M + M_i$$

Applying mass action to this equation gives eqn. 4.58: $[M_M]$ does not appear since to a high degree of accuracy it is simply the number of M sites per unit volume. This is a constant, which is included in the K term.

Combining eqns. 4.59 and 4.58 with 4.57 gives the result

$$[M_i]\,[X_i] = K_7 \tag{4.60}$$

Hence a relationship for interstitials analogous to that for vacancies (eqn. 4.57) is found. Again if some of the interstitial atoms are charged, eqn. 4.60 is correct for the concentrations of uncharged interstitials. For many crystals, both interstitial atoms and vacancies will be important. All of the above equations must then be considered, and solved simultaneously.

4.5.1 Charged defects in compounds

Many native defects (i.e. defects present in the crystal without any addition of foreign atoms) are likely to be charged in compound crystals, although quantitative information is often lacking. The commonest compound semiconductors have either group II or group III atoms on the M site of the lattice. Such an atom has either two or three electrons in its outer shell, so it seems reasonable to assume that if if becomes interstitial it can give up at least one of these to the conduction band. This is just another way of saying that the defect M_i gives rise to an energy level in the forbidden-gap, and acts as an electron donor. The extent to which the donors give up their electrons is determined by the position of the donor level relative to the bottom of the conduction band (Fig. 4.1) and by the location of the Fermi level. The quasichemical description of the donation is

$$M_i \rightleftharpoons M_i^+ + e$$

For interstitial X ions, the reverse is true. The atom is electronegative in character and likely to be an electron acceptor. If this is so, the relevant reaction is

$$X_i \rightleftharpoons X_i^- + h$$

Again it depends on the precise location of the acceptor level on the energy-band diagram as to whether $[X_i^-]$ is an appreciable fraction of $[X_i]$. If $E_F - E_A$ is greater than about $4kT$, all the interstitials are ionised, to a good approximation. The mass-action equations for these last two reactions are

$$K_8 [M_i] = n [M_i^+]$$

and

$$K_9 [X_i] = p [X_i^-]$$

hence

$$K_8 K_9 [M_i] [X_i] = np [M_i^+] [X_i^-]$$

Comparison with eqns. 4.60 and 4.44 gives

$$[M_i^+] [X_i^-] = K_{10} \qquad (4.61)$$

So we have the interesting result that just as the product of neutral M and X interstitials is constant at constant temperature, so is the product of charged interstitials. The equation for electrical neutrality must also be followed, of course, i.e.

$$[M_i^+] + p = [X_i^-] + n$$

The role of vacancies in compound semiconductors is still uncertain. In ionic substances, the metal vacancies are acceptors and the nonmetal ones are donors. This often seems to be the case in II–VI semiconductors as well, but there is very little evidence as to whether or not vacancies are charged in the III–V's. A similar argument to that leading to eqn. 4.61 gives a similar result, i.e.

$$[V_x^+] [V_M^-] = K_{11} \qquad (4.62)$$

To calculate the various defect densities in a given case the procedure is to write down the relevant mass-action equations, together with the neutrality condition, and solve them simultaneously. If a variety of defects is involved this can be an irksome process and a graphical technique due to Brouwer (1954)* becomes useful. Take as a simple example a compound containing as major defects vacancies on both sites. Suppose the M vacancy acts as an acceptor and the X vacancy as a donor. Further assume that the ionisation energies are small, so that they are nearly all charged. The following mass-action relations are relevant. For the reaction between vapour and solid phases

$$[V_X^+]n = K_{12}P_M \tag{4.63}$$

For the electrons and holes

$$np = n_i^2 \tag{4.44}$$

and for the native disorder we have already seen that

$$[V_X^+][V_M^-] = K_{11} \tag{4.62}$$

The neutrality condition is

$$[V_X^+] + p = [V_M^-] + n \tag{4.64}$$

The equilibrium constants K_{11}, K_{12}, n_i^2 are functions only of temperature and can, in principle, be found from experiment. If the temperature is fixed, they can therefore be treated as known constants. Fixing the vapour pressure and temperature now defines the system, since we are then left with four equations and four unknowns, namely $[V_M^-]$, $[V_X^+]$, n and p. This must be so, from the phase rule, since a 2-component 2-phase system, i.e. MX solid, MX vapour, has two degrees of freedom. To solve the problem, it is most useful to solve the four equations for a range of values of P_M. The various defect concentrations can then be plotted as functions of P_M.

The graphical technique is employed as follows. If the logarithm is taken of both sides of eqns. 4.63, 4.44 and 4.62, they give rise to linear equations in the logarithms of the unknowns. This is not so for eqn. 4.64 but it usually happens that one term on each side of eqn. 4.64 predominates, so that a simplified neutrality condition can be used. It is then a relatively simple matter to plot the logarithms of the defect concentrations against $K_{12}P_M$. This is done in Fig. 4.7 for a case in which $n_i^2 = 10^{46}\ \mathrm{m}^{-6}$, $K_{11} = 10^{42}\ \mathrm{m}^{-3}$. In region I we have

$$K_{12}P_M \ll K_{11}$$

i.e.

$$n \ll [V_M^-] \tag{4.65}$$

also

$$\frac{np}{[V_x^+][V_M^-]} = 10^4$$

* Brouwer, G.: Philips Research Repts., 1954, 9, p. 336

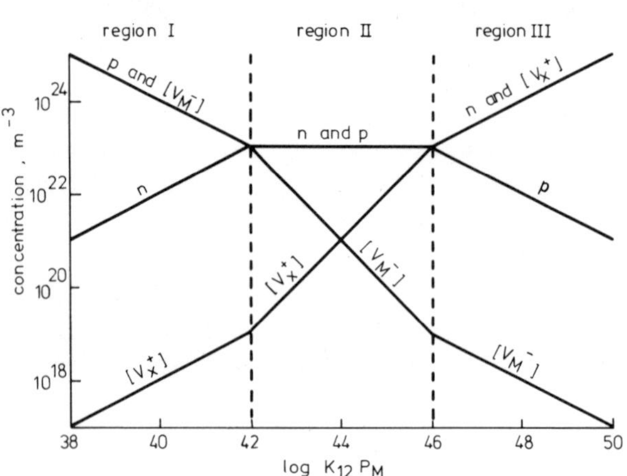

Fig. 4.7 Concentrations of defects in the compound MX

which, in view of eqn. 4.65, indicates that

$$p >> [V_M^-] \tag{4.66}$$

Putting eqns. 4.65 and 4.66 in the neutrality condition shows that in region I,

$$p = [V_M^-] \quad \text{for} \quad \log K_{12}P_M < 42$$

We are now left with three straightforward equations, the solutions of which give the concentrations in region I. In much the same way, it can be shown that in regions II and III the following hold:

$$\textit{region II} \quad n = p \qquad \text{for} \quad 42 < \log K_{12}P_M < 46$$

$$\textit{region III} \quad n = [V_x^+] \qquad \text{for} \quad \log K_{12}P_M > 46$$

The solution is obviously an approximation at the boundaries between the regions, and a better solution is obtained by rounding off the corners. The Figure brings out the physical situation quite nicely. The approximation for region III, for instance, simplifies eqn. 4.63 to $[V_X^+]^2 \propto P_M$, so the line on the Figure for n and $[V_X^+]$ has a slope of 1/2. This graphical technique has been widely used in the interpretation of experiments on II–VI compounds: there seems to be no reason why it should not find more general application.

4.5.2 Foreign atoms

When foreign atoms F are incorporated into the lattice, the basic rules outlined above still apply. The foreign atoms may or may not act as donors or acceptors and the fraction charged depends on the position of an impurity level with respect to the Fermi level. The atom might enter the *MX* lattice as F_M, F_X or F_i. The effect of a substitutional foreign atom depends very largely on the number of valence electrons the atom has, compared to the native atom it is replacing. It is generally true that F acts as a donor if it has more than the native atom, an acceptor if it has less. Thus in CdS, the atoms Ga_{Cd} and Br_S both act as donors, whereas zinc substitutes on the group III site of a III–V semiconductor to give an acceptor level. Usually, no donor or acceptor levels are formed in the forbidden gap if the foreign atom is similar to the one it is replacing, e.g. if aluminium occupies the gallium site in GaAs. Small metal atoms often dissolve interstitially in semiconductors. Usually they behave as donors, ionising to give a positive ion and an electron. In this condition they often have large diffusion coefficients. Copper, for instance, plays this role in a number of semiconductors.

Foreign atoms can enter the crystal either during the original crystal growth process or afterwards from some external phase. It is obviously the latter process which is the more relevant to diffusion work, when the external phase is usually a vapour. If the atom F substitutes for the native atom M in the lattice from an external vapour the reaction is

$$F(\text{vapour}) + V_M \rightleftharpoons F_M$$

and

$$\frac{[F_M]}{[V_M]} = K_{13} P_F$$

where P_F is the external vapour pressure of F. It will be remembered that the concentration of M vacancies depends on the external vapour pressure of M. It therefore follows that the solubility of the neutral F species in *MX* depends on the vapour pressure both of F and of M in the external phase. If the impurity acts as a donor there is a further equation for the charged atoms

$$\frac{n[F_M{}^+]}{[F_M]} = K_{14}$$

If a crystal contains charged native defects as well as foreign donors and acceptors, the situation can become very complicated. The graphical technique described above can be very useful in such a situation. Again it is necessary to divide the conditions into regions in which a relatively small number of defect types is dominant (preferably just two). A diagram similar to Fig. 4.7 can then be drawn, although such diagrams can be a good deal more complex than this.

Fig. 4.7 is drawn for a specific temperature. If we change this temperature, the positions of the boundaries between the regions change, because the mass-action constants are altered. Hence, a system which is dominated by a certain defect at one temperature might be dominated by a different defect at some other temperature. An interesting example of a change in the neutrality condition due to a change in major defect with temperature is shown in the work of Shaw and Whelan (1969)*. They measured the conductivity of CdS as a function of ambient cadmium vapour pressure at a variety of temperatures. The material was n-type. Fig. 4.8 shows a set of results in which the concentration of electrons n is plotted against P_{Cd}.

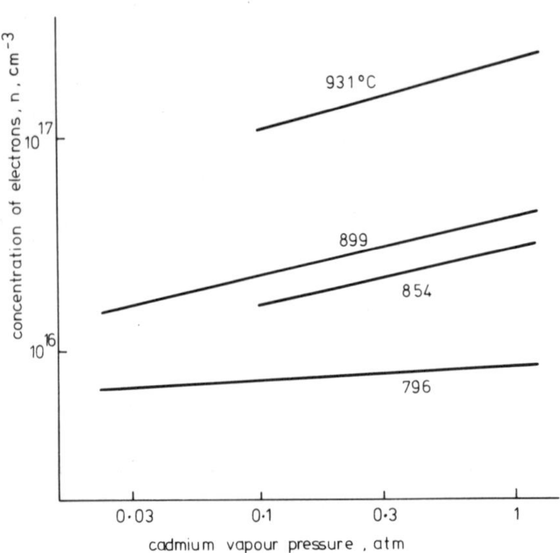

Fig. 4.8 Electron concentration in CdS as a function of ambient cadmium pressure

Reproduced from Shaw, D., and Whelan, R. C.: *Phys. Stat. Sol.,* 1969, **36**, p. 705

Two things are apparent from the Figure. The first is that the pressure of cadmium in the external phase has a marked effect on the number of conduction electrons. The second is that the magnitude of the effect varies with temperature: the results show a variation given by $n \propto P_{Cd}^{\,j}$, where j varies from 0·34 at 931°C to not much more than zero at 796°C. These results were explained by assuming that over the temperature range covered by the experiments there were three main charged defects taking part in the neutrality condition. These were electrons, doubly charged cadmium interstitials Cd_i^{++} and donors of unknown origin, D^+. It was assumed that the donors were all ionised so that $[D^+]$ was always equal to the total number of donors, and therefore had a constant value. The neutrality condition covering all the experiments was therefore

$$n = [D^+] + 2[Cd_i^{++}] \tag{4.67}$$

* Shaw, D., and Whelan, R. C.: Phys. Stat. Solid, 1969, **36**, p. 705

Now the number of cadmium interstitials must increase with temperature. At a sufficiently high temperature, the interstitial concentration will be large compared to that of the donor. The neutrality condition is then

$$n = 2[Cd_i^{++}]$$

A further relation between these quantities can be found from the reaction

$$Cd(\text{vapour}) \rightleftharpoons Cd_i^{++} + 2n$$

so that

$$[Cd_i^{++}] = \frac{KP_{Cd}}{n^2}$$

Substituting this in eqn. 4.67 gives

$$n = (2KP_{Cd})^{1/3}$$

which is the power law observed at 931°C in Fig. 4.8. As the temperature is reduced, the other term in eqn. 4.67 predominates and the neutrality condition becomes

$$n = [D^+] = \text{constant}$$

which is the result to which the curves of Fig. 4.8 are clearly tending as the temperature is reduced.

Chapter 5

Theory of the diffusion coefficient

So far we have adopted a macroscopic view of the phenomenon of diffusion in the sense that, although we have noted the fact that a variation in concentration tends to iron itself out, we have not so far considered what this means on an atomic scale. It is clear, however, that concentrations in a solid can only change by individual atoms moving about within the solid, and it must be possible to relate these movements to the diffusion coefficient D. In this chapter we first of all enquire a little more closely into the idea of diffusion coefficient, and then use random-walk theory to derive an expression for diffusion coefficient for the case in which successive movements of the atoms are random. Finally we consider the way in which these expressions are changed when there is some correlation between successive atom jumps.

5.1 Chemical potential gradient

In previous chapters, two different ways have been described in which one can define equilibrium for a species, and it is appropriate at this point to see how they line up with each other. It has been stated that, in the absence of external fields, a species in a crystal is in equilibrium if there is no concentration gradient in that species. If a concentration gradient does exist, then the atoms move in such a way as to remove it. This leads to the concept of a diffusion coefficient D defined by eqn. 1.1

$$J = -D\frac{\partial C}{\partial x}$$

It has also been stated in a more general context that a species is in equilibrium within a system if the chemical potential of the species is the same everywhere within the system.

Consider a solute diffusing in some solvent and, for the sake of simplicity, confine the problem to one dimension. Suppose the concentration within the solvent at time t is not uniform, and is given by $C(x)$. The chemical potential μ is also a function of distance x. We can treat μ like any other potential and take the view that a gradient in μ must represent a force $\mathscr{F}$ on the atoms, i.e.

$$\mathscr{F} = -\frac{d\mu}{dx} \tag{5.1}$$

When atoms in a crystal are subjected to a force they do not accelerate indefinitely, because of the continual interaction with the other atoms in the lattice. They very rapidly achieve a terminal velocity v that is proportional to the force: the constant of proportionality is called the mobility B'. Hence

$$v = \mathscr{F}B' = -B'\frac{d\mu}{dx}$$

The flux of atoms in the x direction is therefore given by

$$J = Cv = -B'C\frac{d\mu}{dx} \tag{5.2}$$

This equation can now be compared to eqn. 1.1, giving rise to

$$D = B'C\frac{d\mu}{dC} = B'\frac{d\mu}{d\ln C} \tag{5.3}$$

Now μ is normally defined in terms of mole fraction N rather than concentration, and it is therefore convenient to express eqn. 5.3 in this way. This is easy: $d \ln C = d \ln N$, since C and N are proportional to each other. So we have

$$D = B'\frac{d\mu}{d\ln N} \tag{5.4}$$

Chemical potential is defined by eqn. 3.75

$$\mu = \mu_0 + RT \ln \gamma N$$

Substituting this into eqn. 5.4 gives the required result

$$D = B'RT\left(1 + \frac{d\ln\gamma}{d\ln N}\right) \tag{5.5}$$

For a dilute solution, the second term in the bracket is zero, because γ is a constant. Call the value of D for a dilute solution D^*. Eqn. 5.5 then becomes

$$D = D^*\left(1 + \frac{d\ln\gamma}{d\ln N}\right) \tag{5.6}$$

This is an important result. It shows that if the solution of the diffusing species in the crystal goes beyond the Henry's law region, the diffusion constant is affected. Following the nomenclature of Manning (1968)† D^* will be called the tracer-diffusion coefficient. Very often, in semiconductor diffusion, the diffusing species is in very low concentration within the semiconductor. Hence it is D^* that is usually determined in a semiconductor diffusion experiment.

In some experiments there is a concentration gradient of a radioactive isotope of the solute but not of the solute itself. The situation commonly occurs in self diffusion experiments in which solvent and solute are the same element. What is measured is the diffusion coefficient of the radioactive isotope in the crystal, and all that is involved is an interchange of radioactive and nonradioactive atoms. As far as the crystal is concerned, however, nothing is happening at all, since it cannot distinguish between different isotopes of the same atom. The activity coefficient depends only on the total concentration of the element, which is constant, so the second term in the bracket of eqn. 5.6 is zero. The diffusion coefficient measured in a self diffusion experiment is therefore D^*. The same applies to any experiment in which there is an isotopic concentration gradient but not a net concentration gradient of the diffusing species: an example will be given in Chapter 7 when the isoconcentration technique is described.

5.1.1 Interdiffusion coefficients

If the solid solution formed during the diffusion process cannot be described by Henry's law, then the relevant diffusion coefficient is not D^*. The situation then becomes a little more complicated and eqn. 5.6 has to be used. It is sad but true, that what seems to be the simplest example of a diffusion process comes into this category. If a semi-infinite rod, i.e. long compared to a diffusion length, of material A is joined at the plane $x = 0$ to a similar rod of material B, and they are raised to diffusion temperature, an intermingling of atoms takes place. Such an arrangement is usually called a diffusion couple. After diffusion has taken place for some time, the concentration of A varies continuously from almost 100 percent on one side of the couple at $x \ll 0$, to about 0 percent at $x \gg 0$. A similar statement may be made about B. It is obvious that the solution cannot be considered dilute except for a very small part of this range, and the diffusion coefficient will vary according to eqn. 5.6.

A second, quite different, problem arises when considering the diffusion couple. If the diffusion process is allowed to proceed for some time and the concentration of A is then plotted as a function of x, then a diffusion coefficient can be assigned to the profile. If the same is done for substance B, the same diffusion coefficient will be found, since, at all points in the couple, $N_A = 1 - N_B$. The coefficient is therefore a measure of the way in which the two types of atoms mix. There is no reason, however, why the individual coefficients D_A and D_B should be equal, and it

† Manning, J. R., 'Diffusion kinetics for atoms in crystals' (Van Nostrand, 1968)

is obviously of some interest to consider the way in which the coefficients D, D_A and D_B are related.

Let the lengths of the 'semi-infinite' rods be l and choose a set of co-ordinates in which the rod of material A occupies $0 < x < l$ and that of B occupies $l < x < 2l$, as shown in Fig. 5.1. Introduce a marker of some sort to mark the plane at $x = 0$. It is not necessary to choose this particular plane, any plane in the diffusion region would do. Before diffusion starts, there is an equal volume of material on each side of the reference plane. If we suppose that $D_A > D_B$, the flux of atoms from left to right J_A will be greater than that from right to left J_B once the diffusion process starts. This means that the volume to the right of the reference plane will increase, and that to the left will shrink during diffusion. The marker will therefore move steadily towards the end of the couple at $x = 0$ throughout the experiment. The phenomenon has, in fact, been observed in metals and is usually called the Kirkendall effect.

Consider a marked plane M and suppose that during the course of diffusion it moves with velocity v_M with respect to the ends of the couple. Define the x axis so that the left-hand end of the couple is always at $x = 0$ (Fig. 5.1). An observer who is stationary with respect to the marked plane would see a flux of A given by

$$-D_A \frac{\partial C_A}{\partial x}$$

A second observer, who is stationary on the x co-ordinate system would see the marked plane (and the first observer) moving at v_M. He would therefore measure the flow of A as

$$J_A = -D_A \frac{\partial C_A}{\partial x} + v_M C_A \tag{5.7}$$

Similarly he would give the flow of B as

$$J_B = -D_B \frac{\partial C_B}{\partial x} + v_M C_B$$

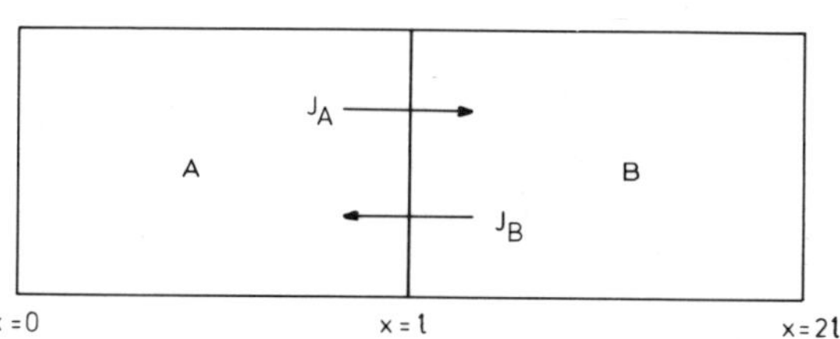

Fig. 5.1 Diffusion between two 'semi-infinite' rods

The net flow, according to the second observer is therefore

$$J = J_A + J_B = v_M C - D_A \frac{\partial C_A}{\partial x} - D_B \frac{\partial C_B}{\partial x} \tag{5.8}$$

where $C = C_A + C_B$ and represents the total concentration of atoms.

Now consider what the second observer, stationary with respect to the x axis, actually observes during this experiment. The net flow of atoms he sees is zero. If he started off with half of the bar to his left and half to his right, this is still the case at the end of the experiment (we shall ignore any effect due to equal numbers of A and B atoms occupying slightly different volumes). All that has changed is the composition, i.e. percentages of A and B in the portions of the bar to his left and right. He would therefore measure no net flow of atoms, i.e. $J = 0$. Putting this in eqn. 5.8 gives

$$v_M = \frac{1}{C}\left(D_A \frac{\partial C_A}{\partial x} + D_B \frac{\partial C_B}{\partial x}\right)$$

$$= \frac{1}{C}\left(D_A \frac{\partial C_A}{\partial x} - D_B \frac{\partial C_A}{\partial x}\right) \tag{5.9}$$

where the substitution $\partial C_A/\partial x = -\,\partial C_B/\partial x$ has been made.

Substituting for v_M in eqn. 5.7 gives

$$J_A = -\frac{\partial C_A}{\partial x}\left(\frac{C_A D_B + C_B D_A}{C}\right)$$

This is simply Fick's law, with the term in brackets giving the effective diffusion coefficient D. It is convenient to express it in terms of mole fractions N_A, N_B.

$$D = N_A D_B + N_B D_A \tag{5.10}$$

Similarly the velocity of the marker can be given as

$$v_M = (D_A - D_B)\frac{\partial N_A}{\partial x} \tag{5.11}$$

Eqns. 5.10 and 5.11 express the situation quite nicely. If the diffusion coefficients of the two components in a diffusion couple are equal, then the interdiffusion coefficient also has this value. If they are different, however, the interdiffusion coefficient is a weighted mean, and any marker in the lattice will move with a velocity given by eqn. 5.11.

It is important to remember that over almost all of the length of the couple the mixture of A and B is not a dilute solution. It follows that the diffusion coefficients in eqn. 5.10 are not the tracer coefficients and D_A, D_B are given by eqn. 5.6 as

$$D_A = D_A^* \left(1 + \frac{\partial \ln \gamma_A}{\partial \ln N_A}\right)$$

$$D_B = D_B^* \left(1 + \frac{\partial \ln \gamma_B}{\partial \ln N_B}\right) \tag{5.12}$$

The two activity coefficients can be compared using the Gibbs–Duhem relation (eqn. 3.27).

$$N_A \, d\mu_A + N_B \, d\mu_B = 0 \tag{5.13}$$

From the definition of chemical potential

$$N_A \, d\mu_A = RT(dN_A + N_A \, d \ln \gamma_A) \tag{5.14}$$

with a similar equation for $N_B \, d\mu_B$. Substituting these into eqn. 5.13 yields

$$dN_A + N_A \, d \ln \gamma_A + dN_B + N_B \, d \ln \gamma_B = 0$$

Dividing by dN_A and noting that $dN_A = -dN_B$ gives

$$\frac{d \ln \gamma_A}{d \ln N_A} = \frac{d \ln \gamma_B}{d \ln N_B}$$

The two sets of brackets in eqn. 5.12 are therefore equal and either can be used to substitute in eqn. 5.10. Taking the first, eqn. 5.10 becomes

$$D = (N_A D_B^* + N_B D_A^*) \left(1 + \frac{d \ln \gamma_A}{d \ln N_A}\right) \tag{5.15}$$

This equation has been checked in a number of cases for metals, and has been found correct within experimental error.

5.2 Atomic mechanisms

Having looked at the idea of a diffusion coefficient, we now consider the way in which individual atoms can move to bring about diffusion. Which of them actually operates in a given situation can only be decided by experiment.

5.2.1 Vacancy mechanisms

We have already seen that at equilibrium any crystal contains a certain number of vacancies, given by eqn. 4.8. At any temperature above absolute zero, all of the atoms are in fairly violent oscillatory motion, and from Fig. 5.2 it can be seen, that if the tracer atom in position 1 is moving in the right direction at the right time, it can move into the vacant space at position 4. Or to put it another way, the atom

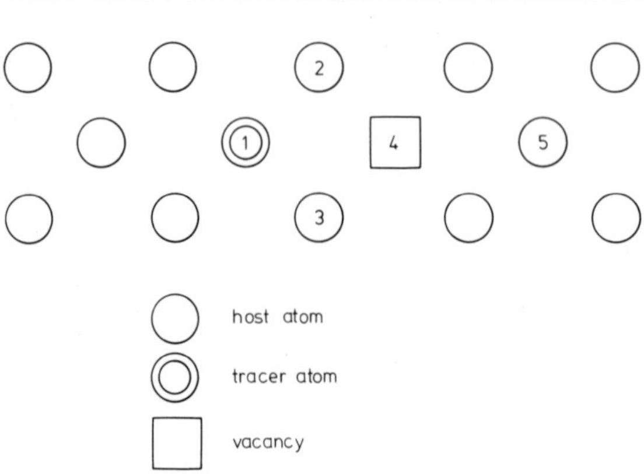

Fig. 5.2 Diffusion by a vacancy mechanism

and the vacancy change places. The vacancies also move through the crystal by this mechanism, so that subsequently one may appear in position 5 and the tracer atom may make a further move to the right. In this way it can diffuse through the crystal.

It is clear that there must be an energy barrier to overcome in order for the tracer to make the move 1–4. The normal lattice positions must, by definition, be the positions of energy minima for atoms in the crystal. For an atom to leave 1 and go to 4 it must leave a minimum, push its way over a potential hill (which by symmetry, has its maximum value halfway between sites) and run down the hill to the next site position. If we adopt the simple view of a crystal as being made up of close-packed spheres, then it can be seen that the tracer atom cannot change places with the vacancy unless the atoms at 2 and 3 separate a little in order to let it through. (In fact, when considered in three dimensions, more than two atoms have to move, since atoms in planes above and below that shown also move.) The activation energy can be thought of in this model as the energy required to push atoms 2 and 3 apart by the necessary amount. An atom which reaches the midpoint between 1 and 4, and is also moving to the right, will diffuse. A configuration with an atom in this position is called an activated complex, and it is of some interest to calculate how many such complexes are likely to occur in a crystal. This will be considered later in the chapter.

In crystals in which there is a strong binding energy between vacancies, diffusion can take place by way of divacancy pairs. The mechanism is very similar to that described above, although it is not identical because of the lack of symmetry of a divacancy. Normally diffusion by single vacancies predominates over any divacancy mechanism.

5.2.2 Interstitial mechanisms

When an atom exists in the lattice on interstitial sites, it can diffuse by hopping from one such site to another, as shown in Fig. 5.3. The situation is very similar to that for vacancy diffusion, with one important exception. With a vacancy mechanism, a diffusing atom is usually surrounded by atoms on the nearest-neighbour sites, and only rarely does a vacancy arrive to help it diffuse. In a dilute interstitial solution, however, a diffusing atom is virtually always surrounded by empty interstitial sites, and so it does not have to wait for an adjacent site to become empty. This is, of course, a great help to the diffusion process. On the other hand, there is usually less room at an interstitial site, so interstitial diffusion is likely to be favoured by small atoms in relatively open lattices. Apart from this, similar conditions apply to those for vacancy mechanisms. In Fig. 5.3 it can be seen that atoms 3 and 4 must move to allow the diffusing atom to go from position 1 to position 2, and an activated complex can be defined as the situation in which the atom is half-way between positions 1 and 2.

If the diffusing interstitial atom is not very small, it is obviously going to have difficulty in pushing its way through to the next interstitial site. Diffusion is then more likely to occur by the interstitialcy mechanism, in which the manoeuvre is executed in two steps rather than one. The mechanism is illustrated in Fig. 5.4. The interstitial atom is originally at site 1. Instead of moving to an adjacent interstitial site, such as 4, it goes to the lattice site, 2, pushing the resident atom to site 3. The interstitialcy has then moved half a lattice spacing. In principle, the lattice atom might be pushed to a site such as 4 rather than 3, but this is much less likely.

The interstitial mechanism can be further generalised by stating that interstitial-type diffusion can take place in a crystal region in which there is one more atom than lattice site. This is illustrated in Fig. 5.5, in which it may be seen that the middle row contains nine atoms, whereas the other rows have only eight. The configuration of the middle row is called a 'crowdion'. It is symmetrical about the dotted line drawn halfway between columns 4 and 5, and it is convenient to say

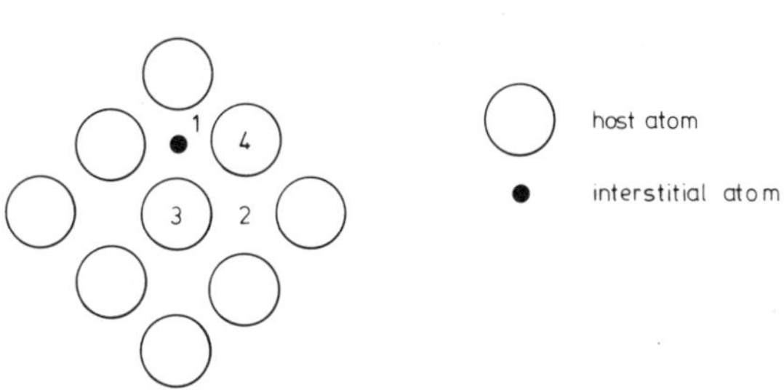

Fig. 5.3 Diffusion by an interstitial mechanism

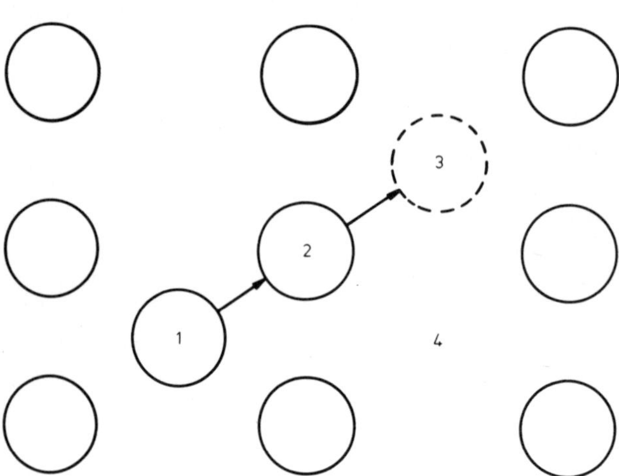

Fig. 5.4 Intersticialcy mechanism

that the crowdion is centred on this position. The atoms in the middle row are displaced only slightly from proper lattice position. It takes only a very small movement of each atom to the right for the crowdion to move so that it is centred between columns 5 and 6. In this way it can move all along the row: in passing from one end of the row to the other, the crowdion effectively moves the atoms by one lattice spacing. This manner of moving adopted by the crowdion is very similar to the way in which an edge dislocation moves through a crystal.

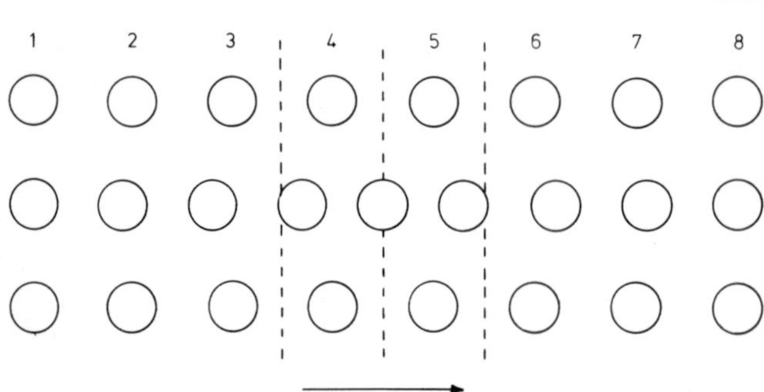

Fig. 5.5 Crowdion

5.2.3 Exchange mechanisms

The simplest way for an atom to move through a crystal lattice would be for it to change places with an adjacent atom. Such a mechanism has the attractive feature that no lattice defects are required for it to operate. However, the lattice distortion would be large and for this reason it would be unlikely to happen. One way in which the mechanism could operate with less distortion would be for more than two atoms to be involved. A ring mechanism involving four atoms is shown in Fig. 5.6. The four atoms rotate in a clockwise direction and the tracer atom is moved one space to the right. This mechanism seems less likely than the substitutional and interstitial ones mentioned above, and there is little evidence that it occurs.

5.3 Random-walk diffusion

It is apparent that whichever of the above mechanisms operates in a given crystal, an atom finds its way through the lattice by consecutive jumps from one site to the next. Jumps between nearest neighbours are generally so much more probable than others, that it is only necessary to consider these. If the jumps are random with no preferred direction, the atom is said to be following a 'random walk'. Crystal diffusion is rather simple in this respect: the jumps can have only certain discrete lengths, corresponding to the unit cell of the crystal, and they can be made only in certain directions. We need only to consider random-walk diffusion by the vacancy mechanism, since the ordinary interstitial mechanism can be considered as a special case in which the diffusing atom is always completely surrounded by vacancies. We start with a simple treatment of diffusion in one dimension, which tells us what physical parameters are important in defining the value of D.

Take as an example the body-centred cubic lattice shown in Fig. 5.7a. Suppose that diffusion takes place by a vacancy mechanism along a [100] direction, which can be taken as the x axis. The length of the cube edge is a. It can be seen from the Figure that an atom has eight nearest neighbours, of which four are on plane 3 and

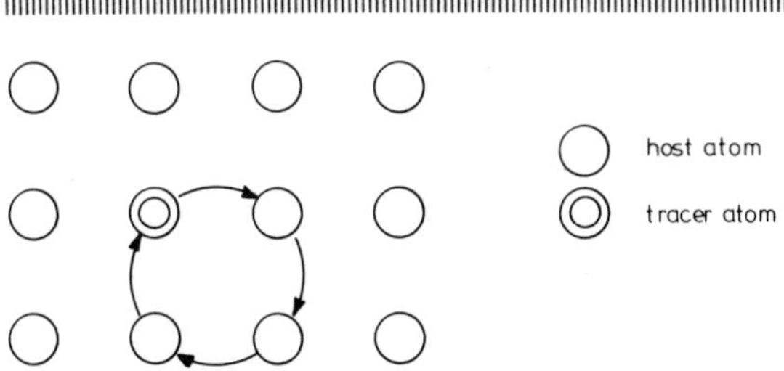

Fig. 5.6 Ring mechanism of diffusion

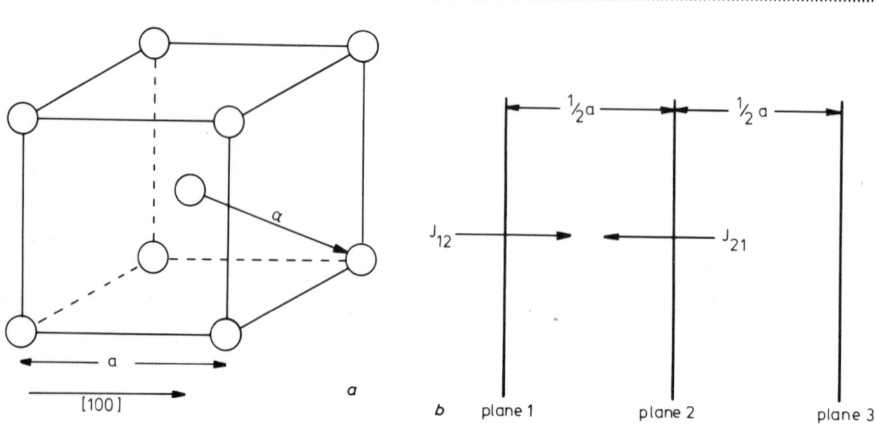

Fig. 5.7 *a* Elemental jump in a body-centred cubic lattice

b Diffusion flux between adjacent (100) planes

four are on plane 1 (Fig. 5.7b). The jump distance, labelled α in the Figure is given by $\alpha = a\sqrt{3}/2$. Suppose the frequency with which an atom jumps is ν. Since an atom has equal numbers of nearest neighbours to the left and to the right, it follows that half of these jumps are in the positive x direction and half are in the negative x direction. The flux from plane 1 to plane 2 can therefore be written

$$J_1 = n_1 \cdot \tfrac{1}{2}\nu$$

where n_1 is the concentration per unit area of the diffusing atoms. A similar equation can be written for the flux between planes 2 and 1, so that the net flux of atoms from 1 to 2 is

$$J = J_1 - J_2 = \tfrac{1}{2}\nu(n_1 - n_2)$$

Now the concentrations per unit area, n_1, n_2, are related to those per unit volume by the relation

$$n = \tfrac{1}{2}aC$$

so that

$$J = \tfrac{1}{4}a\nu(C_1 - C_2) \qquad (5.16)$$

The term in brackets may be expressed:

$$\frac{C_1 - C_2}{\tfrac{1}{2}a} = \frac{\partial C}{\partial x}$$

which, on substituting into eqn. 5.16 yields

$$J = \frac{1}{8}\, a^2 \nu \,\frac{\partial C}{\partial x}$$

It is more convenient to express this equation using the jump distance α, rather than the lattice constant a. Using the relation given above

$$J = \frac{1}{6}\, \alpha^2 \nu \,\frac{\partial C}{\partial x}$$

This is Fick's law, with the diffusion coefficient given by

$$D = \tfrac{1}{6}\alpha^2 \nu \tag{5.17}$$

We could have chosen any of the cubic lattices and still got the same equation (eqn. 5.17). Here D is the 'tracer' coefficient defined above: the asterisk notation is used only when there is some danger of confusion.

Diffusion in a [100] direction was considered in this derivation. The result is correct for any crystal orientation, however, since the diffusion coefficient in a cubic crystal is a scalar, having the same value in any direction (Chapter 2).

It is worth looking more closely at the jump frequency ν. The number of jumps in time δt is $\nu\delta t$. Suppose an atom has a vacancy on a nearest-neighbour site in a b.c.c. crystal. Let the probability that it will jump into it in time δt be $\omega\delta t$ (note that ω is a frequency). The total number of jumps in δt must be $\omega\delta t$ times the number of available vacancies on nearest-neighbour sites. The latter quantity is $8P_v$, where P_v is the probability of a lattice site containing a vacancy. This argument gives

$$\nu\delta t = 8P_v\omega\delta t$$

Now, numerically, the probability P_v must be equal to the mole fraction of vacancies N_v. Hence $\nu = 8N_v\omega$, giving

$$D = \tfrac{4}{3}\alpha^2 N_v \omega \tag{5.18}$$

A similar argument can be carried through for interstitial diffusion, but it is simpler because in a dilute interstitial solution, the mole fraction of vacant sites is essentially unity.

5.3.1 Temperature dependence of *D*

It was mentioned in Chapter 1 that diffusion coefficients frequently show a variation with temperature of the form

$$D = D_0 \exp\frac{-Q}{RT} \tag{1.3}$$

From eqn. 5.18 it would appear that this temperature dependence should be a consequence of the behaviour of the parameters ω and N_v. The latter has been considered in some detail in Chapter 4. The equilibrium mole fraction of vacancies in a crystal was found to be

$$N_v = \exp\frac{\Delta S_v}{R}\exp\frac{-\Delta H_v}{RT} \tag{4.8}$$

A thermodynamic treatment has been given by Wert (1950)* in which he considers the behaviour of the frequency ω. The essentials of the argument are as follows. As already mentioned, an atom must climb over a potential barrier to pass from one site to another. When it is in the region of the maximum value, it is called an activated complex. Suppose there are at any one time n_m such complexes. The number of atoms diffusing, per unit time, is given by the number of complexes divided by the time taken to cross the barrier t. The number of atoms diffusing per second is therefore $n_m v/\delta$, where v is the mean velocity of atoms moving between sites and δ is the width of the complex. The average jump frequency per atom ω is then given by

$$\omega = \frac{N_m v}{\delta}$$

where N_m is the mole fraction of atoms in activated complexes. Now let us assume that an activated complex can be treated just like any other crystal defect. The Gibb's free energy of an atom in a complex is greater than that of an atom on a lattice site by an amount ΔG_m, say. If we can apply the same reasoning that was applied to the other defects, then an equation of the form of eqn. 4.8 applies for N_m.

$$N_m = \exp\left(\frac{-\Delta G_m}{RT}\right)$$

Hence ω has the form

$$\omega = \omega_0 \exp\frac{-\Delta G_m}{RT} \tag{5.19}$$

And it is usually assumed that ω_0 is of the order of magnitude of the vibrational frequency of an atom about its mean position. Substituting eqns. 5.19 and 4.8 into eqn. 5.18 gives an expression for the diffusion coefficient in a b.c.c. lattice:

$$D = \frac{4}{3}\alpha^2\omega_0 \exp\left(\frac{\Delta S_v + \Delta S_m}{R}\right)\exp\left(\frac{-\Delta H_v - \Delta H_m}{RT}\right) \tag{5.20}$$

* Wert, C. A., *Phys. Rev.*, 1950, **79**, p. 601

which is of the form of eqn. 1.3, with

$$D_0 = \frac{4}{3} \alpha^2 \omega_0 \exp\left(\frac{\Delta S_v + \Delta S_m}{R}\right)$$

and

$$Q = (\Delta H_v + \Delta H_m)$$

5.3.2 Length of random walk

We now consider the length of a random walk, and relate this to the diffusion coefficient. Suppose an atom takes n random jumps, given by the vectors $\mathbf{r}_1$, $\mathbf{r}_2$, $\mathbf{r}_3 \ldots, \mathbf{r}_n$. Let the vector joining the starting and finishing points be $\mathbf{R}$. The mean value of $\mathbf{R}$ for many such sets is designated $\langle\mathbf{R}\rangle$ and the mean-square value $\langle R^2\rangle$. Now the first of these quantities must be zero, since if many atoms take n jumps, they will end up evenly distributed about the origin. The vector sum of the $\mathbf{R}$'s is therefore zero. $\langle R^2\rangle$ is not in general zero, however and this will be calculated.

For a given set of n jumps, where n is a large number, we have

$$\mathbf{R} = \mathbf{r}_1 + \mathbf{r}_2 + \mathbf{r}_3 + \ldots + \mathbf{r}_n$$

giving the vector dot product as

$$\begin{aligned} R^2 = \mathbf{R} \cdot \mathbf{R} = & \mathbf{r}_1 \cdot \mathbf{r}_1 + \mathbf{r}_1 \cdot \mathbf{r}_2 + \ldots + \mathbf{r}_1 \cdot \mathbf{r}_n \\ & + \mathbf{r}_2 \cdot \mathbf{r}_1 + \mathbf{r}_2 \cdot \mathbf{r}_2 + \ldots + \mathbf{r}_2 \cdot \mathbf{r}_n \\ & \vdots \\ & + \mathbf{r}_n \cdot \mathbf{r}_1 + \mathbf{r}_n \cdot \mathbf{r}_2 + \ldots + \mathbf{r}_n \cdot \mathbf{r}_n \end{aligned} \tag{5.21}$$

It is convenient to divide the terms into two; the diagonal ones of the type $\mathbf{r}_i \cdot \mathbf{r}_i = r_i^2$, and the off diagonals, of the form $\mathbf{r}_i \cdot \mathbf{r}_{i+j}$. Since $\mathbf{r}_i \cdot \mathbf{r}_{i+j} = \mathbf{r}_{i+j} \cdot \mathbf{r}_i$, the terms above the diagonal are equal to those below and eqn. 5.21 simplifies to

$$R^2 = \sum_{i=1}^{n} \mathbf{r}_i^2 + 2\sum_{i=1}^{n-1} \mathbf{r}_i \cdot \mathbf{r}_{i+1} + 2\sum_{i=1}^{n-2} \mathbf{r}_i \cdot \mathbf{r}_{i+2} + \ldots$$

$$R^2 = \sum_{i=1}^{n} \mathbf{r}_i^2 + 2\sum_{j=1}^{n-1}\sum_{j=1}^{n-j} \mathbf{r}_i \cdot \mathbf{r}_{i+j} \tag{5.22}$$

In a cubic crystal, all jump vectors to nearest-neighbour sites have the same length α. We can note further that if $\theta_{i,i+j}$ represents the angle between the ith and $(i+j)$th jumps, then $\mathbf{r}_i \cdot \mathbf{r}_{i+j} = \alpha^2 \cos\theta_{i,i+j}$ and eqn. 5.22 becomes

$$R^2 = n\alpha^2 + 2\alpha^2 \sum_{j=1}^{n-1}\sum_{i=1}^{n-j} \cos\theta_{i,\, i+j} \tag{5.23}$$

To find the mean-square displacement $\langle R^2 \rangle$, it is necessary to average R^2 over many trajectories of n jumps. The first term in eqn. 5.23 will be the same for every trajectory. Since the movements are random, jumps in any allowed direction are equally probable. Positive and negative values of any given $\cos\theta$ term will therefore appear in equal numbers, and the mean value of the second term in eqn. 5.23 will be zero. For a random walk, therefore, one obtains the simple expression

$$\langle R^2 \rangle = n\alpha^2 \tag{5.24}$$

As an example of eqn. 5.24, consider an atom diffusing in a crystal in 2Å jumps. Suppose it makes 10^8 jumps per second and it is kept at diffusion temperature for 10^5 seconds (rather more than one day). Eqn. 5.24 gives the root-mean-square diffusion distance as 6×10^{-2} cm. If, on the other hand, all of its jumps were in one straight line, the distance would be 2 km.

If the jumps are not completely random, i.e. if successive jumps are correlated in some way, the second term in eqn. 5.23 cannot be ignored. This situation will be considered below. A comparison of eqns. 5.17 and 5.24 gives the relation between mean-square displacement and diffusion coefficient. Bearing in mind the fact that the time to make n jumps is given by $t = n/\nu$, the equation for diffusion coefficient becomes

$$D = \frac{\langle R^2 \rangle}{6t} \tag{5.25}$$

5.4 Correlation effects

The main assumption made in considering random-walk diffusion is that atom jump probabilities do not depend on the directions of preceding jumps. In real crystals, however, this is not usually the case, and jump probabilities do depend on the previous movements of the atom. Take as an example the tracer atom diffusing in a simple two dimensional lattice such as that of Fig. 5.2. If a tracer diffuses by a vacancy mechanism, then it can move only if a vacancy comes to an adjacent site, as in the Figure. Suppose that the last move of the tracer was $4 \rightarrow 1$. The next move is clearly not random, since it is more likely to return to site 4 than to move to any of the other five nearest neighbours. This is because it can move directly to 4, whereas for it to move to site 3, for example, the vacancy must first exchange with 3 and then with 1. There is another way in which it might conceivably move to 3: a second vacancy might come to that site. The concentration of vacancies in a crystal is very low, however, and the probability of this happening is so small that it can safely be ignored.

If we regard the vacancy in Fig. 5.2 as the diffusing species, then we can see that its movements are random. All six nearest-neighbour sites are always occupied by atoms, and there is no preferred direction for the vacancy to jump (strictly, this is true only if the tracer atom is a radioactive isotope of the host atom, rather than a different element.) In the same way, the jumps of an interstitial atom in Fig. 5.3 are uncorrelated because the diffusing atom is always surrounded by empty sites, providing the interstitial solution is dilute. Therefore no correlation effects arise when the environment is symmetrical, as for a vacancy in a pure metal or an interstital in a dilute solution. If the tracer element in Fig. 5.2 were a different element to the host atoms the jumps of the vacancy would become correlated since the probability of exchange with the tracer would be different to that of exchange with a host atom. For this reason, we confine ourselves for the present to self diffusion of a radioactive tracer by a vacancy mechanism.

This tendency for a tracer atom to return to the site whence it has just come is obviously a bit counter-productive from the diffusion point of view, and has the effect of changing the 'random' diffusion coefficient of eqn. 5.17 to the 'correlated' coefficient

$$D_c = \tfrac{1}{6}\alpha^2 \nu f \tag{5.26}$$

where f is a constant, equal to unity for random-walk diffusion. We can write

$$f = \frac{D_C}{D} = \frac{\langle R_C{}^2 \rangle}{\langle R^2 \rangle}$$

where the second equality comes from eqn. 5.25. $\langle R_C{}^2 \rangle$ is the mean-square distance for a correlated walk and $\langle R^2 \rangle$ is for a random walk consisting of the same number of steps. $R_C{}^2$ is given by the expression eqn. 5.23 and R^2 by its simplified version, (eqn. 5.24). Hence

$$f = 1 + \left\langle \frac{2}{n} \sum_{j=1}^{n-1} \sum_{i=1}^{n-j} \cos\theta_{i,\,i+j} \right\rangle \tag{5.27}$$

where, once again, the averaging is carried out over a large number of trajectories, each consisting of n jumps. Mathematically speaking, there is no difference between averaging the sum of the $\cos\theta$ terms and doing the reverse, i.e. summing the averages. Eqn. 5.27 can therefore just as well be written

$$f = 1 + \frac{2}{n} \sum_{j=1}^{n-1} \sum_{i=1}^{n-j} \langle \cos\theta_{i,\,i+j} \rangle \tag{5.28}$$

Consider the meaning of $\langle \cos\theta_{i,i+j} \rangle$. It is the average value, for a lot of flights, of the angle between the ith jump and the $(i+j)$th jump (more precisely the mean cosine of that angle). Now there is no reason why this should depend on i. The mean value of $\cos\theta$ should depend just on how many jumps separate the two vectors so, for instance, one would expect

$$\langle \cos\theta_{1,1+j} \rangle = \langle \cos\theta_{2,2+j} \rangle = \langle \cos\theta_{3,3+j} \rangle \text{ etc.} \tag{5.29}$$

The above equation is therefore more complicated than it need be and we can write

$$f = 1 + \frac{2}{n}\sum_{j=1}^{n-1}(n-j)\langle\cos\theta_{i,\ i+j}\rangle \tag{5.30}$$

The i notation is, strictly speaking, now redundant, but it is convenient to hang on to it for a few more lines of working. It has already been noted that n has to be a large number for the statistics to be acceptable. We can therefore at this stage let n tend to infinity. Eqn. 5.30 then gives a value of f

$$f = 1 + 2\langle\cos\theta_{i,i+1}\rangle + 2\langle\cos\theta_{i,i+2}\rangle + \ldots \tag{5.31}$$

Fortunately we can simplify even further, because it turns out that the cos θ terms in eqn. 5.31 are related by a simple equation, providing the crystal demonstrates at least a 2-fold symmetry. The equation therefore applies to crystals of cubic symmetry and close-packed hexagonal. Consider the ith, $(i + 1)$th and $(i + 2)$th jumps. The first and second vectors define one plane and the second and third define another. If the angle between these planes is β, there is a geometrical relationship (Campaan and Haven, 1956)*

$$\cos\theta_{i,i+2} = \cos\theta_{i,i+1}\cos\theta_{i+1,i+2} + \sin\theta_{i,i+1}\sin\theta_{i+1,i+2}\cos\beta \tag{5.32}$$

If every jump vector lies along an axis either of 2-fold or of 3-fold symmetry, then for every angle β that appears, an angle $(\pi - \beta)$ also appears, with equal probability. Since $\cos\beta = -\cos(\pi - \beta)$, this means that when expression 5.32 is averaged, the second term comes to zero and the equation gives

$$\langle\cos\theta_{i,i+2}\rangle = \langle\cos\theta_{i,i+1}\cos\theta_{i+1,i+2}\rangle \tag{5.33}$$

Comparing eqn. 5.33 with eqn. 5.29 and the argument that preceded it gives

$$\langle\cos\theta_{i,i+2}\rangle = \langle\cos\theta_{i,i+1}\rangle^2$$

and, by iteration

$$\langle\cos\theta_{i,i+k}\rangle = \langle\cos\theta_{i,i+1}\rangle^k$$

We now drop the 'i' part of the notation and write for f:

$$f = 1 + 2\langle\cos\theta_1\rangle + 2\langle\cos\theta_1\rangle^2 + 2\langle\cos\theta_1\rangle^3 + \cdots$$

and $\langle\cos\theta_1\rangle$ is the mean value for the angle between consecutive jumps. This is a convergent series, with the sum

$$f = \frac{1 + \langle\cos\theta_1\rangle}{1 - \langle\cos\theta_1\rangle} \tag{5.34}$$

Calculating f therefore comes down to finding a value for $\langle\cos\theta_1\rangle$.

* Campaan, K, and Haven, Y., *Trans. Faraday Soc*, 1956, **52**, p. 786

5.4.1 Evaluation of $\langle \cos \theta_1 \rangle$

Finding $\langle \cos \theta_1 \rangle$ and then f can become very complicated. To give the reader a feel for what is involved, a very simple example is given for the 2-dimensional square lattice of Fig. 5.8.

The last move taken by the tracer was $f \rightarrow g$, and the jump vector is that shown in Fig. 5.8. The vacancy therefore moved in the opposite direction $g \rightarrow f$. We wish to find the average angle for the next jump. Assume that the concentration of vacancies is low, so that the only way the tracer can change its position is by interaction with the vacancy now at f. Only nearest neighbour jumps are allowed, so an atom cannot move diagonally on the diagram. The fact that the jumps of the vacancy are random provides us with the means of carrying out the calculation.

The next jump of the tracer will be to f, c, h or k. Let the probabilities be P_f, P_c, P_h, P_k respectively and call the angles these jumps make with the previous jump θ_f, θ_c, θ_h, θ_k. Then

$$\langle \cos \theta_1 \rangle = P_f \cos \theta_f + P_c \cos \theta_c + P_h \cos \theta_h + P_k \cos \theta_k$$

In this simple case

$$\cos \theta_c = \cos \theta_k = 0$$

$$\cos \theta_f = -1$$

$$\cos \theta_h = 1$$

So

$$\langle \cos \theta_1 \rangle = P_h - P_f$$

First consider P_f. The obvious way for the first move of the tracer to be to f would be for it to change places immediately with the vacancy. From the point of view of the vacancy, this is just as likely as exchanging with any of its other three nearest neighbours, so the probability is ¼. This exchange is carried out with only one movement of the vacancy. There is no reason why the vacancy should not move

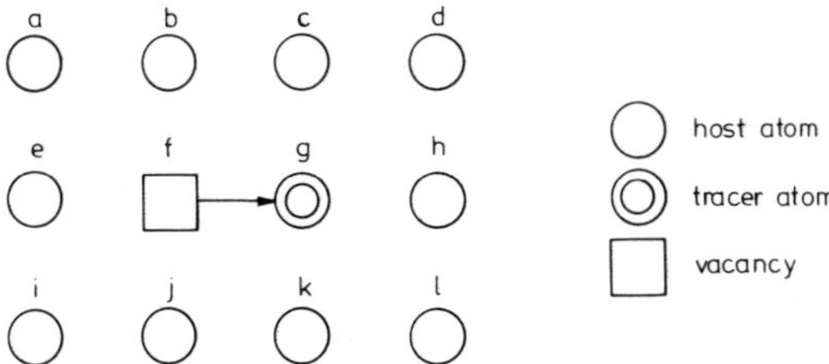

Fig. 5.8 Correlation effects

more than once before exchanging with the tracer, however, and the possibility must be considered. There is no way that the required exchange can be made in two moves of the vacancy, but there are three ways in which it can be made in three moves. These are the (vacancy) trajectories *fbfg, fefg, fifg*. All of these paths give the desired effect of moving the tracer from g to f on its first move. Since three moves of the vacancy are involved and the probability of a single move is ¼, the probability of each of the trajectories is $(\frac{1}{4})^3$. There are again no ways in which the vacancy can carry out its mission in four moves. P_f is therefore given by a series, the first few terms of which are

$$P_f = \tfrac{1}{4} + 0 + 3(\tfrac{1}{4})^3 + 0 + 16(\tfrac{1}{4})^5 + \cdots$$

The series converges, but quite slowly, since although the term $(\frac{1}{4})^n$ gets small rapidly as n gets larger, the number of ways the vacancy can bring about the desired effect gets quite large with increasing n. We shall draw the line at five jumps and take the above equation as an approximation.

Applying similar reasoning to P_h, we find five jumps of the vacancy is the minimum required to take the tracer to site h on its first move. There are two such paths, *fbcdhg* and *fjklhg*. Since we are not going beyond five jumps, the approximate value for P_h is

$$P_h = 2(\tfrac{1}{4})^5$$

Using these values for P_f and P_h

$$\langle \cos\theta_1 \rangle \simeq 0{\cdot}002 - 0{\cdot}313 = -0{\cdot}311$$

This compares with the exact figure of $-0{\cdot}363$ given by Campaan and Haven (1958).* The inaccuracy in the above is due to our having taken the series no further than the fifth term.

It is now a simple matter to find f. The exact value for this lattice, and a number of others calculated by Campaan and Haven (1958)* are given in Table 5.1.

This technique, whilst being quite instructive in showing the basic principles involved, is hardly practical in dealing with more difficult examples: real 3-dimensional crystals, for instance. Even in the calculation given above, the method became unwieldy before the series had properly converged. A variety of mathematical techniques have been devised to calculate $\langle \cos\theta_1 \rangle$ and f, including the use of electrical analogue circuits. The reader is referred to an article by Le Claire (1970).†

5.4.2 Impurity diffusion by vacancies

So far we have dealt only with correlation effects in self-diffusion. When the tracer atom is different from the host atom, a new dimension is added to the correlation

* Campaan, K., and Haven, Y.: *Trans. Faraday Soc.*, 1958, **54**, p. 1498

† Le Claire, A. D.: Chap. 5 *in* 'Physical chemistry – an advanced treatise. Vol. 10' (Ed. W. Jost) (Academic Press, 1970)

Table 5.1 Correlation factors for self diffusion

(from Campaan, K., and Haven, Y.: *Trans. Faraday Soc.*, 1958, **54**, 1498)

2-dimensional square lattice	0·46694
2-dimensional hexagonal lattice	0·56006
Diamond	0·50000
Simple cubic	0·65311
Body-centred cubic	0·72722
Face-centred cubic	0·78146
Hexagonal close-packed	0·78121 (normal to c axis 0·78146 (parallel to c axis

effects which are possible, and the situation becomes very much more complicated. Only a brief indication of what is involved can be given here.

Consider initially the case of a 2-dimensional-square lattice so that Fig. 5.8 still suffices, this time with the tracer as a foreign element. We still have

$$\langle \cos \theta_1 \rangle = P_h - P_f$$

and since we have found $|P_f| >> |P_h|$, the first approximation we might make is

$$\langle \cos \theta_1 \rangle \simeq - P_f$$

The first term in the series for P_f is the probability that the vacancy at f will go to g, changing places with the tracer atom. This probability is no longer ¼, because the probability of exchanging with the tracer is no longer the same as the probability that the vacancy will change places with its other nearest neighbours. Let ω_t be the jump frequency for exchanging with a tracer, and ω_s be the frequency for exchanging with a solvent atom. Then the first term in the series becomes

$$\frac{\omega_t}{\omega_t + 3\omega_s}$$

If we make the further, rather rash, assumption that P_f can be approximated by the first term in the series, then eqn. 5.34 gives the correlation coefficient.

$$f \simeq \frac{3\omega_s}{2\omega_t + 3\omega_s}$$

The very rough nature of the approximations can be seen by reducing the example back to the self-diffusion case. The two jump frequencies are then equal again, and the approximate correlation factor becomes 0·6. This is to be compared to the exact value of 0·46694 given in Table 5.1.

As a matter of fact, the approximation becomes slightly less awful as the number of nearest neighbours is increased. A lattice site in a simple cubic crystal has six nearest neighbours, for instance. Using the above argument again, in which we only consider the return of the tracer to its original site, the correlation factor becomes

$$f \simeq \frac{5\omega_s}{2\omega_t + 5\omega_s}$$

Again putting $\omega_s = \omega_t$, f becomes 0·714, compared to the 0·65311 given in Table 5.1.

Some improvement can be made in the approximations, however. In the above treatment, only the return of the tracer to its original site is considered in calculating $\langle \cos \theta_1 \rangle$ and f. This amounts to saying that the tracer will move only if the first move of the vacancy is to the impurity site, i.e. $f \to g$ in Fig. 5.8. Any other move by the vacancy is considered a loss, as far as impurity movement is concerned. This is clearly not correct. Again referring to Fig. 5.8, a move $f \to e$ by the vacancy is not necessarily a loss to the impurity if its next move is to return from e back to f. The 'escape frequency' of the vacancy from the impurity area is therefore not $3\omega_s$ (for a square lattice), but rather less than this. Call it $3f'\omega_s$, where $0 < f' < 1$. The equation for correlation coefficient for the 2-dimensional square lattice becomes

$$f = \frac{3f'\omega_s}{2\omega_t + 3f'\omega_s}$$

and similarly for the simple cubic lattice,

$$f = \frac{5f'\omega_s}{2\omega_t + 5f'\omega_s}$$

The simple way to obtain a value for f' is to take the case for self diffusion. The two frequencies are then equal, and all that is necessary is to substitute in one of the above expressions the accurate value of f taken from Table 5.1. Hence for the simple cubic lattice,

$$f = 0.653 = \frac{5f'}{2 + 5f'}$$

giving $f' = 0{\cdot}754$. With a more complex set of assumptions, f' can become a variable: details are given in Manning's book.

5.4.3 Effect on diffusion coefficient

A proper understanding of the way in which atoms move about in crystals cannot be achieved without taking account of correlation effects. However, it must also be borne in mind that f usually turns out to be less than, but of the order of, unity. The values of the other quantities making up D in eqn. 5.26 are by no means accurately known, generally speaking, and when a single determination of D is made experimentally, it is often quite in order to ignore f completely.

When a series of values of D is taken as a function of temperature, the presence of f can be significant, however. It has been noted above that a relationship is commonly found of the form

$$D = D_0 \exp \frac{-Q}{RT}$$

An activation energy can therefore be found from

$$Q = -R \frac{d \ln D}{d\left(\frac{1}{T}\right)}$$

In the absence of correlation effects, the measured activation energy Q can be put equal to the enthalpies of formation of the defects involved ΔH say. Now if the correlation effects are to be included, the equation should more properly be written

$$D = f D_0 \exp\left(\frac{-\Delta H}{RT}\right)$$

and ΔH and Q may not now coincide. If f is temperature dependent, the measured activation energy is then

$$Q = -R \frac{d \ln D}{d\left(\frac{1}{T}\right)} = \Delta H - R \frac{d \ln f}{d\left(\frac{1}{T}\right)}$$

so that the value of Q does not necessarily correspond to the enthalpy term. Whether or not this is a significant effect depends, of course, on the magnitude of the temperature variation of f. For impurity diffusion by a vacancy mechanism, f depends on the relative values of the various jump frequencies. As the temperature varies, the ratios between the frequencies also vary and, in principle, this could give rise to a large temperature dependence in f. For self diffusion in pure cubic crystals all the jump frequencies are the same, so f is a constant independent of temperature and $Q = \Delta H$.

Almost all of the experimental work studying correlation effects in crystals has been carried out on metals and ionic compounds. Very little seems to be known to date about the importance of the effects in semiconductors. It is probably true to say that the detailed study of correlation effects represents a level of sophistication that has not yet been reached in semiconductor diffusion work. This is not the same thing as saying that such effects are of no importance in semiconductors, however.

Chapter 6

Diffusion mechanisms in semiconductors

The reader might be forgiven at this point for thinking that most of the story has been told. Providing the solid solution formed by the diffusion process is dilute, the system can be described by a constant diffusion coefficient, D, and it is merely a matter of applying the boundary conditions appropriate to the experiment in hand, and solving Fick's law, as was done in Chapter 2. Would that it were so! In fact, experimental results frequently do not compare well with the mathematical results of Chapter 2, and one is forced to the conclusion that it is often improper to regard D as a simple constant. This is not to say, however, that simple interstitial or substitutional diffusion processes never occur in semiconductors: they just do not occur very often in a straight-forward manner. It seems very likely, for instance, that lithium diffuses interstitially in both silicon and germanium, and many self diffusion processes are substitutional. In this chapter we start with an account of the extra complications involved in substitutional and interstitial mechanisms when the semiconductor is a compound rather than an element. This is followed by descriptions of a number of mechanisms that have been proposed to account for nonsimple diffusion behaviour in semiconductors. Finally a fairly full account is given of one of the most important of the 'complex' diffusion mechanisms: the substitutional–interstitial mode.

6.1 Substitutional and interstitial mechanisms in compounds

It was shown in the previous chapter that a substitutional mechanism gives rise to a diffusion coefficient that is proportional to vacancy concentration. This is quite unambiguous for an elemental semiconductor in which the vacancy concentration depends only on the temperature but, as pointed out in Chapter 4, this is not so for a compound. For the compound MX, the product of the two types of vacancies is constant at constant temperature, as given by eqn. 4.57 i.e.

$$[V_M]\,[V_X] = K_1 \qquad (4.57)$$

Consider an atom occupying the M site in the compound. A simple substitutional mechanism requires the atom to exchange with a vacancy on the nearest M site. Now M sites are not in general nearest-neighbour sites, but next-nearest neighbours on the lattice. Thus the basic diffusion jump is much less likely than for substitutional diffusion in an element, in which the basic jump is between adjacent sites. The mechanism can therefore be expected to be much less common in compounds and, when it does occur, to be characterised by a rather small diffusion coefficient. When it occurs, the diffusion coefficient will obviously depend critically on the appropriate vacancy concentration, here $[V_M]$, which can be fixed by fixing the vapour pressure of either the M or X component surrounding the crystal. Suppose we decide to fix the vapour pressure of X and that it exists as the dimer X_2. The reaction describing the equilibrium between lattice and vapour is

$$X_2 \text{ (gas)} + 2V_X \rightleftharpoons 2X_X$$

giving the mass-action relation

$$P_{X_2} [V_X]^2 = K_2$$

which, combined with eqn. 4.57 leads to the result

$$D_M \propto [V_M] \propto P_{X_2}{}^{1/2} \tag{6.1}$$

for a substitutional diffusion mechanism. One way of testing to see if such a mechanism operates, is therefore to observe the way in which the coefficient varies as a function of ambient vapour pressure.

A similar situation occurs for interstitial diffusion in a compound. Consider, for instance, self diffusion of element M in the compound MX and suppose that a small proportion of the element occurs in the interstitial form. M can then diffuse either substitutionally or interstitially. We consider parallel diffusion modes more carefully in Section 6.3, but it may perhaps be stated here as being intuitively reasonable that if the interstitial diffusion coefficient is so much larger than the substitutional one that the latter can be ignored, then the overall coefficient will be proportional to the fraction of M that is interstitial, i.e. $D \propto [M_i]$.

Now the two interstitial concentrations are related by eqn. 4.60

$$[M_i]\,[X_i] = K_3 \tag{4.60}$$

and it was also shown in Chapter 4 that

$$[V_M]\,[M_i] = K_4 \tag{4.58}$$

$$[V_M]^2 = K_5 P_{X_2} \tag{4.55}$$

where again the example is taken of X appearing in the vapour phase as a dimer. Eliminating $[V_M]$ from these last two equations gives

$$D \propto [M_i] \propto P_{X_2}{}^{-1/2} \tag{6.2}$$

This relationship is the inverse of the pressure dependance for a substitutional mechanism. This type of variation with ambient pressure has been observed in the self diffusion of Cd in CdS (Shaw and Whelan, 1969)* (although only under certain conditions of external vapour pressure), leading to the assumption that the diffusing species was Cd_i

6.2 Built-in field effect (see also Section 2.2.8)

When an ionised donor or acceptor diffuses in the presence of an electric field $\mathscr{E}$ Fick's law has to be modified to take account of the field. For an ionised donor, concentration C, the diffusion equation becomes

$$J = -D_D \frac{\partial C}{\partial x} \pm B_D \mathscr{E} C \tag{6.3}$$

D_D and B_D are the diffusion coefficient and mobility of the ionised donor, related by the Einstein equation

$$\frac{D_D}{B_D} = \frac{kT}{e}$$

and a singly ionised donor is assumed. The sign of $\mathscr{E}$ in eqn. 6.3 depends on whether the imposed field is assisting the diffusion or retarding it.

It is now shown that a diffusing ionised donor (or acceptor) normally does diffuse in an electric field which is essentially of its own making. When the donor first entered the lattice, it did so as a neutral atom. It then split into a positively ionised donor and an electron and the two components were left to diffuse separately. Now the mobility of an electron in the lattice B_n is very much greater than that of the donor. The electrons therefore get ahead of the donor, on the average, and a space-charge is set up. The resulting field is such as to help the motion of the donors and hinder that of the electrons. One might expect, therefore, that the effect of this built-in field is to bring about an increase in diffusion coefficient and this is indeed so. In the following rather simple treatment of the effect, it is assumed that all donors are ionised. A rather more sophisticated version has been given by Klein and Beale (1966)† in which they consider both donors and acceptors diffusing simultaneously.

There are three sets of charged particles involved in this problem, namely ionised donors, electrons and holes. They each give rise to an electric current due to the

* Shaw, D., and Whelan, R. S.: *Phys. Stat. Solid*, 1969, **36**, p. 705

† Klein, T., and Beale, J. R. A., *Solid-State Electron.*, 1966, 9, p. 59.

effects of normal diffusion and the electric field. Call these currents j_D, j_n, j_p. They are given by

$$j_D = -eD_D \frac{\partial C}{\partial x} + eB_D C\mathscr{E} \tag{6.4}$$

$$j_n = eD_n \frac{\partial n}{\partial x} + eB_n n\mathscr{E} \tag{6.5}$$

$$j_p = -eD_p \frac{\partial p}{\partial x} + eB_p p\mathscr{E} \tag{6.6}$$

The space-charge neutrality condition gives

$$n = C + p \tag{6.7}$$

The use of eqn. 6.7 might seem a little odd, in view of the fact that it has been explicitly stated above that the space-charge does exist and, indeed, gives rise to the effect under discussion. However, it can be shown (Lehovic and Slobodsky, 1961)* that the magnitude of the spacecharge is small compared to C, so that while eqn. 6.7 is not strictly valid, it is still a good approximation.

There can be no net current flow in the crystal, so the sum of the currents must equal zero

$$j_D + j_n + j_p = 0 \tag{6.8}$$

Substituting in eqn. 6.8 from eqns. 6.4 to 6.6 and solving for $\mathscr{E}$ gives

$$\mathscr{E} = \frac{kT}{e} \cdot \frac{B_D \frac{\partial C}{\partial x} - B_n \frac{\partial n}{\partial x} + B_p \frac{\partial p}{\partial x}}{B_D C + B_n n + B_p p} \tag{6.9}$$

The effect will be largest when the concentration of diffusing donors is large (it is obviously nonexistent when C = 0), and it is sufficient for our purposes to consider the maximum possible effect of the built-in field. This will occur when the crystal is strongly *n*-type so that eqn. 6.7 becomes

$$n \simeq C \quad \text{and} \quad p \simeq 0$$

The equation for E becomes

$$\mathscr{E} = \frac{kT}{e} \cdot \frac{\frac{\partial C}{\partial x}(B_D - B_n)}{C(B_D + B_n)}$$

* Lehovic, K., and Slobodsky, A., *Solid-State Electron.,* 1961, 3, p. 45.

which, since $B_n >> B_D$ simplifies to

$$\mathscr{E} = -\frac{kT}{e} \cdot \frac{1}{C} \cdot \frac{\partial C}{\partial x} \tag{6.10}$$

so the built-in field is proportional to the concentration gradient. The flux of donor atoms is given by the modified diffusion equation (eqn. 6.3).

$$J = -D_D \frac{\partial C}{\partial x} - \frac{kT}{e} \cdot B_D \frac{\partial C}{\partial x}$$

where the positive sign has been used from eqn. 6.3, since the effect of the field is to aid diffusion. Replacing B_D with D_D by Einstein's relation gives the result

$$J = -2D_D \frac{\partial C}{\partial x} \tag{6.11}$$

So the maximum effect of the built-in field is to double the effective diffusion coefficient. In many mechanisms in which D varies, this factor of two (or less) is small enough to be ignored. The built-in field effect will not be allowed for explicitly in any of the other mechanisms to be described in this chapter, although it will often be present as a further effect.

6.3 Diffusion of partially ionised impurities

One of the more common complications appears when the diffusing atom occurs in more than one form. Call these forms 1 and 2, so that the concentrations of the two species at a point x are $C_1(x)$ and $C_2(x)$. We then have

$$C_1 + C_2 = C$$

where C is the total concentration. The diffusion flux of the atoms is then the sum of the fluxes due to the two species, i.e.

$$J = -D_1 \frac{\partial C_1}{\partial x} - D_2 \frac{\partial C_2}{\partial x} \tag{6.12}$$

This can be rewritten

$$J = -D_1 \left(\frac{\partial C_1}{\partial C} + D_2 \frac{\partial C_2}{\partial C} \right) \frac{\partial C}{\partial x} \tag{6.13}$$

and the term in brackets is the effective diffusion coefficient. In the special case of C_1 and C_2 always being in the same proportions; $\partial C_1/\partial C = C_1/C$ and $\partial C_2/\partial C = C_2/C$, and the effective diffusion coefficient is

$$D = \frac{D_1 C_1}{C} + \frac{D_2 C_2}{C} \tag{6.14}$$

If C_1 and C_2 are more complicated functions of C, however, eqn. 6.13 must be used to determine the diffusion flux.

When a donor or acceptor impurity occurs in a semiconductor it can exist in either the ionised or unionised state. An acceptor, for instance, is negatively charged if it has accepted an electron from the valence band and neutral if it has not. The proportions of neutral and ionised acceptors are not related simply, but are given by Fermi-Dirac statistics. Eqn. 6.14 does not, therefore, apply. Let us take the acceptor as an example, a case which has been considered in some detail by Allen (1960).* Let the total concentration of acceptors be C, the ionised concentration C^- and the ionised C^0. Then

$$C = C^- + C^0 \tag{6.15}$$

The number ionised is given by the Fermi-Dirac expression

$$C^- = \frac{C}{1 + \frac{1}{2}\exp\left(\frac{E_I - E_F}{kT}\right)} \tag{6.16}$$

where E_I is the energy level of the impurity and E_F is the Fermi energy. If the acceptors are reasonably shallow (i.e. if the impurity levels are quite close to the valence band) and if this concentration is much greater than n_i, the electrons play a negligible part in fixing the Fermi level and the condition of charge neutrality becomes

$$C^- = p \tag{6.17}$$

The number of holes is given by eqn. 4.24

$$p = 2\left(\frac{2\pi m_h kT}{h^2}\right)^{3/2} \exp\left(\frac{E_V - E_F}{kT}\right)$$

Eqns. 6.15 to 6.17 and eqn. 4.24 give sufficient information to determine C^- and C^0 as functions of C. They cannot be solved explicitly, but Allen carried out a numerical calculation to determine the fraction ionised. He used parameters which were at the time considered reasonable for GaAs, namely an effective mass for electrons of 0·04 m (m being the normal electron mass, $9{\cdot}1 \times 10^{-31}$ kg), an effective mass for holes of 0·4 m and a shallow acceptor level 0·1 eV above the valence band. The result of his calculation is shown in Fig. 6.1, in which the fraction ionised is plotted against total concentration. At low concentrations virtually all acceptors are ionised, but as the semiconductor becomes degenerate and the Fermi level enters the valence band the proportion drops quite sharply.

In general, the impurity atom will have different diffusion coefficients in the

* Allen, J. W., *J. Phys. & Chem. Solids,* 1960, **15,** p. 134

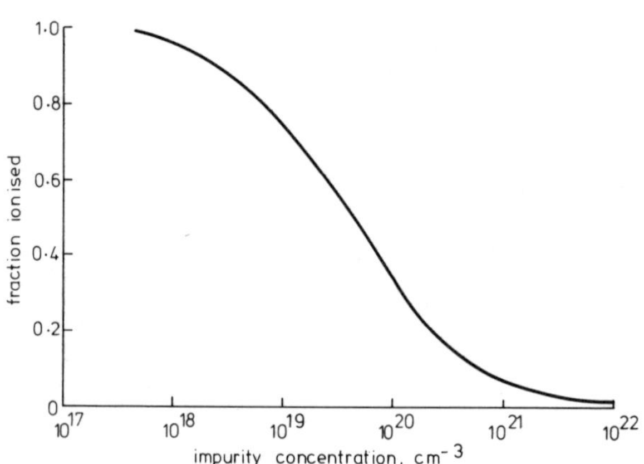

Fig. 6.1 Fraction of acceptors ionised as a function of impurity concentration

Reproduced from Allen, J. W.: *J. Phys. & Chem. Solids*, 1960, **15**, 134

two states. Call these D^+ and D^0. Using eqn. 6.12, it can be seen that the diffusion profile is a solution of Fick's law

$$\frac{\partial}{\partial x}\left(D^+\frac{\partial C^+}{\partial x}+D^\circ\frac{\partial C^\circ}{\partial x}\right)=\frac{\partial C}{\partial t} \tag{6.18}$$

Since C^+ and C^0 are, in principle, known functions of C, this equation can be solved. Once again it can only be done numerically. The problem was simplified by Allen to a form which allowed an explicit solution. He made the approximation that above a certain concentration of impurities, N, say, none of the acceptors have accepted an electron, whereas below N, all of them have. This amounts to simplifying the curve of Fig. 6.1 to a step function. Considering diffusion into a semi-infinite slab $x>0$ with the surface at $x=0$ maintained at the surface concentration C_s, the diffusion equation becomes

$$D^\circ\frac{\partial^2 C}{\partial x^2}=\frac{\partial C}{\partial t} \qquad C>\mathrm{N} \tag{6.19}$$

$$D^+\frac{\partial^2 C}{\partial x^2}=\frac{\partial C}{\partial t} \qquad C<\mathrm{N}$$

with boundary conditions

$$C=0 \quad at \quad t=0 \quad for \quad x>0$$

$$C=C_s \quad for \quad t>0 \quad at \quad x=0$$

There is the further condition that at the change-over concentration, N, the flux of atoms must be continuous, i.e.

$$D^{\circ}\frac{\partial C}{\partial x} = D^{+}\frac{\partial C}{\partial x} \quad at \quad C = \mathrm{N} \tag{6.20}$$

Eqn. 6.20 shows that if $D^0 \neq D^+$ there must be a discontinuity in the slope of the diffusion profile at $C = \mathrm{N}$. For the detailed solution of eqn. 6.19, the reader is referred to the original paper. However it is fairly clear that for concentrations greater than N, the profile is primarily determined by the diffusion coefficient for unionised atoms D^0 and below N, it is dominated by D^+. In Fig. 6.2 a profile is plotted for the following parameters:

$$C_s = 5 \times 10^{20}\ \mathrm{cm}^{-3}, \quad \mathrm{N} = 5 \times 10^{19}\ \mathrm{cm}^{-3}, \quad D^0 = 4D^+, \quad (D^0 t)^{½} = 10^{-2}\ \mathrm{cm}$$

where t is diffusion time. The predicted discontinuity occurs at the concentration $5 \times 10^{19}\ \mathrm{cm}^{-3}$ and the diffusion front drops away very sharply because of the relatively low value of D^+. This mechanism was originally suggested to explain the diffusion of zinc in GaAs, but subsequent work has established fairly conclusively that this system diffuses by a substitutional-interstitial mechanism, to be described later in the chapter.

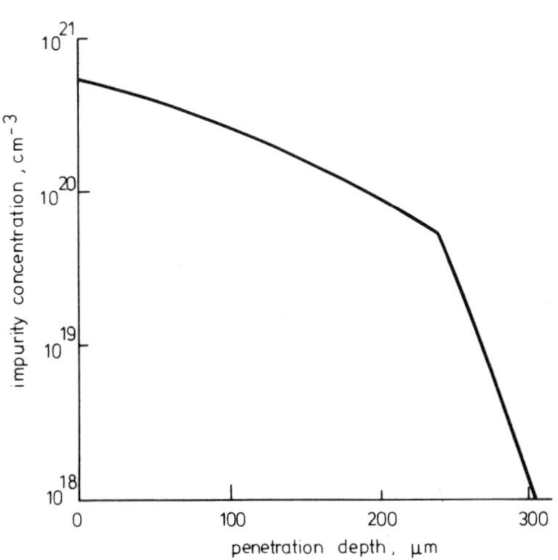

Fig. 6.2 Diffusion profile for an acceptor diffusing by the mechanism proposed by Allen. For concentrations above $5 \times 10^{19}\ \mathrm{cm}^{-3}$ the diffusion is dominated by unionised atoms. Below this concentration it is determined by ionised atoms

Reproduced from Allen, J. W.: *J. Phys. & Chem. Solids,* 1960, **15**, p. 134

6.4 Effects due to charged vacancies

It has already been noted in Chapter 4 that vacancies can act as acceptors in semiconductors and that this can have an effect on the diffusion coefficient. In this section we look at the phenomenon a little more closely.

The total concentration of vacancies is given by

$$[V] = [V^0] + [V^-] \tag{6.21}$$

$[V^0]$ is the concentration of neutral vacancies, and $[V^-]$ represents the ionised variety. To a good approximation, $[V^0]$ is the thermal equilibrium value, which depends only on temperature. The concentration of charged vacancies can be related to $[V]$ by the Fermi–Dirac function

$$[V^-] = \frac{[V]}{1 + \tfrac{1}{2}\exp\left(\dfrac{E_A - F}{kT}\right)} \tag{6.22}$$

In which E_A is the vacancy acceptor level, and F is the Fermi energy. (note the temporary change in notation for Fermi energy. It is to avoid the use of a double subscript.) Eqns. 6.21 and 6.22 give the relation

$$[V^\circ] = \frac{[V]}{1 + 2\exp\dfrac{F - E_A}{kT}} \tag{6.23}$$

Now consider two elemental semiconductors, one doped and the other intrinsic. Let the total concentration of vacancies in the first be $[V]_d$ and that in the second be $[V]_i$ and let F_d, F_i be the respective Fermi levels. Apart from the difference in doping, the semiconductors are to be considered identical. Then from eqn. 6.23 it follows that

$$\frac{[V]_d}{[V]_i} = \frac{1 + 2\exp\left(\dfrac{F_d - E_A}{kT}\right)}{1 + 2\exp\left(\dfrac{F_i - E_A}{kT}\right)} \tag{6.24}$$

The electron concentrations will be different in the two semiconductors: call them n_d, n_i. From eqn. 4.23 we have

$$n_d \propto \exp\frac{F_d - E_c}{kT}$$

$$n_i \propto \exp\frac{F_i - E_c}{kT}$$

so that

$$\frac{n_d}{n_i} = \exp\frac{F_d - F_i}{kT} \tag{6.25}$$

Eqn. 6.24 can therefore be written

$$\frac{[V]_d}{[V]_i} = \frac{1 + 2\dfrac{n_d}{n_i}\exp\left(\dfrac{F_i - E_A}{kT}\right)}{1 + 2\exp\left(\dfrac{F_i - E_A}{kT}\right)} \tag{6.26}$$

It remains to consider how the diffusion constant is affected. In the simplest type of substitutional mechanism, the coefficient is proportional to the vacancy concentration, and we can write

$$\frac{D_d}{D_i} = \frac{[V]_d}{[V]_i}$$

The intrinsic Fermi level will be approximately at the middle of the forbidden gap, so if the vacancy acceptor level is close to the valence band, $F_i - E_A >> kT$ and

$$\frac{D_d}{D_i} = \frac{n_d}{n_i} \tag{6.27}$$

If, however, the acceptor level is close to the centre of the band gap, the exponential term tends to unity and the ratio of diffusion coefficients is

$$\frac{D_d}{D_i} = \frac{1}{3}\left(1 + 2\frac{n_d}{n_i}\right) \tag{6.28}$$

The equations predict that if vacancies act as acceptors in a crystal, any substitutional-type diffusion process should show a larger diffusion coefficient in n-type than in p-type material. The proposition is most conveniently checked experimentally using self diffusion because the diffusing atoms do not dope the crystal. The doping can therefore be fixed before the diffusion experiment and can be assumed constant throughout. The self-diffusion constant for silicon has been measured in this way (Ghoshtagore, 1966).* The experiment was carried out using heavily doped n type, $n = 8{\cdot}0 \times 10^{19}$ cm^{-3}, intrinsic, $n = p = 3{\cdot}1 \times 10^{13}$ cm^{-3} and heavily doped p type, $p = 9{\cdot}5 \times 10^{19}$ cm^{-3}. The self-diffusion coefficients were $7{\cdot}3 \times 10^{-14}$ cm^2 s^{-1}, $4{\cdot}6 \times 10^{-14}$ cm^2 s^{-1} and $3{\cdot}3 \times 10^{-14}$ cm^2 s^{-1} respectively, showing the correct qualitative trend.

If impurity diffusion in this type of system is considered, a further complication occurs because the diffusing species in general adds to the overall doping of the crystal. The Fermi level is no longer simply fixed by the original background doping

* Ghoshtagore, R. N., *Phys. Rev. Lett.*, 1966, **16**, p. 890

and varies with distance within the crystal. The number of ionised vacancies therefore varies also and $[V]_d/[V]_i$ is no longer a constant. This problem can be overcome if it is possible to arrange for the maximum concentration of the diffusing species, i.e. the surface concentration, always to be much less than the original background doping. The diffusing impurity then affects the Fermi level by a negligible amount, and the concentration of vacancies is fixed. This type of experiment has been carried out in silicon by Millea (1966).* He found the behaviour described above for the donors phosphorus and antimony, i.e. as he made the progression from *n* type, to intrinsic, to *p* type, the diffusion coefficients decreased. For two other diffusing elements, indium and tin, he did not find this behaviour. This led to the conclusion that the first two elements diffuse in silicon by a simple substitutional mechanism and the last two do not. Some of his results are shown in more detail in Table 6.1.

If the concentration of the diffusing impurity is comparable to the background doping, then n_d in eqns. 6.26 to 6.28 is no longer a constant, but depends on the concentration of diffusant at any point. If this is a diffusing donor, for instance, and its concentration is much greater than background, then we have $n_d = C$, where C is the concentration of the diffusing donor. The diffusion coefficient then ceases to be a simple constant and becomes a variable dependent on the concentration of the diffusing species. If the vacancy level lies near the valence band, then eqn. 6.27 applies and the diffusion coefficient becomes proportional to concentration.

Table 6.1 Effect of uniform doping on diffusion coefficients of impurities

(from Millea, M. F.: *J. Phys. Chem. Solids*, 1966, **27**, 315)

Tracer	Temperature, K	Type of doping	Diffusion coefficient $cm^2\ s^{-1}$	n_d cm^{-3}	$\frac{D_d}{D_i}$	$\frac{n_d}{n_i}$
P		n	$9{\cdot}2 \times 10^{-12}$	$6{\cdot}3 \times 10^{19}$	1·6	2·0
P	1525	intrinsic	$5{\cdot}9 \times 10^{-12}$	—	—	—
P		p	$4{\cdot}1 \times 10^{-12}$	$5{\cdot}0 \times 10^{18}$	0·7	0·16
Sb		n	$7{\cdot}1 \times 10^{-13}$	$7{\cdot}1 \times 10^{19}$	2·5	3·0
Sb	1465	intrinsic	$2{\cdot}8 \times 10^{-13}$	—	—	—
Sb		p	$8{\cdot}2 \times 10^{-14}$	$2{\cdot}4 \times 10^{18}$	0·29	0·10
In		n	$1{\cdot}8 \times 10^{-12}$	$8{\cdot}1 \times 10^{19}$	0·58	3·1
In	1488	intrinsic	$3{\cdot}1 \times 10^{-12}$	—	—	—
In		p	$2{\cdot}5 \times 10^{-11}$	$3{\cdot}5 \times 10^{18}$	8·1	0·13
Sn		n	$2{\cdot}5 \times 10^{-13}$	$8{\cdot}8 \times 10^{19}$	2·1	3·0
Sn	1488	intrinsic	$1{\cdot}2 \times 10^{-13}$	—	—	—
Sn		p	$1{\cdot}7 \times 10^{-13}$	$3{\cdot}9 \times 10^{18}$	1·4	0·07

* Millea, M. F., *J. Phys. & Chem. Solids*, 1966, **27**, p. 315

When charged vacancies in compounds are considered, the further interesting possibility arises of the two different types of vacancy being oppositely charged. Assume that for a compound MX the vacancies are V_M^- and V_X^+. The probability of divacancies being formed is then much greater, since the two types are attracted by coulombic forces: this is considered in Section 6.7. Suppose, however, that this does not happen to any great extent but the concentrations of vacancies are very large, and exceed the intrinsic carrier concentration at the diffusion temperature n_i. The charge neutrality condition will then be dominated by the vacancy concentrations and we have

$$[V_M^-] = [V_X^+] \tag{6.29}$$

Since the concentrations will also exceed the concentrations of neutral vacancies eqn. 6.29 leads to the conclusion that the numbers of vacancies are constant at constant temperature. We cannot vary the ratio of $[V_M^-]$ to $[V_X^+]$ by manipulating the external vapour pressures, because this would upset eqn. 6.29. A situation then arises in which the self-diffusion coefficients are constant, independent of the external vapour pressures. This mechanism has been suggested for self-diffusion in ZnO.

6.5 The interchange mechanism

In the work described in the previous section, group V donors diffusing in silicon fitted the charged-vacancy theory rather well, approximately obeying eqn. 6.27. The group III acceptor indium did not fit the theory, however. In fact the inverse of eqn. 6.27 was followed fairly accurately, i.e.

$$\frac{D_d}{D_i} = \frac{n_i}{n_d} \tag{6.30}$$

Millea put forward the following suggestion to account for this behaviour. The acceptor level for indium is fairly low in the forbidden gap, i.e. fairly near to valence band. It follows that the indium will occur in an unionised form In^0 and an ionised form In^-, and that the latter will predominate at diffusion temperatures. It must be remembered, however, that the ionised form for the vacancies is V^-, so there must be a coulombic repulsion between the ionised indium and the ionised vacancies. It would be rather surprising, therefore, if the ionised indium could diffuse through the charged vacancies (note that the reverse is true for donors: they will be attracted to the vacancies.) Looked at in this light, it is fairly obvious that acceptors and donors will obey different diffusion laws. Two possibilities arise. In the first the In^- ions diffuse through uncharged vacancies. It has already been pointed out, however, that the concentration $[V^0]$ is a constant depending only on temperature. Such a mechanism would therefore give rise to a constant value for diffusion coefficient, in contradiction to eqn. 6.30.

The other possibility is that the diffusion is carried out by the rather small fraction of indium that is neutral. The mechanism suggested by Millea is that an indium atom leaves a lattice site to move to an interstitial position, leaving behind a vacancy. A neighbouring silicon atom moves into the vacancy and the indium atom moves into the place whence the silicon atom came, thus completing the interchange of adjacent silicon and indium atoms. (Millea suggested that the indium atom actually does this as In^+, the electron diffusing separately to the new position to reconstitute the In^0 atom. The principle is the same, however.)

For this mechanism the diffusion coefficient is proportional to the fraction of indium that is neutral. The fraction depends on Fermi level position and is given by eqn. 6.23

$$\frac{[In^\circ]}{[In]} = \frac{1}{1 + 2\exp\left(\frac{F - E_A}{kT}\right)}$$

where E_A is the indium acceptor level

which, if $F - E_A >> kT$ becomes

$$\frac{[In^\circ]}{[In]} = \tfrac{1}{2} \exp \frac{E_A - F}{kT}$$

So comparing the neutral indium concentrations in a doped and an intrinsic sample, we have

$$\frac{[In^\circ]_d}{[In^\circ]_i} = \frac{D_d}{D_i} = \exp \frac{F_i - F_d}{kT} \tag{6.31}$$

Comparing eqn. 6.31 with eqn. 6.25 shows

$$\frac{D_d}{D_i} = \frac{n_i}{n_d}$$

which is the observed result.

The results of Table 6.1 indicate that the diffusion coefficient of tin in silicon is almost independent of doping. This is also a result to be expected if an interchange mechanism is operative, since substitutional tin, being group IV, is normally neutral in silicon.

The interchange mechanism obviously requires sufficient interstitial space to accommodate the diffusing atom. It might be expected, therefore, that a crystal lattice with large interstitial spacing would demonstrate the mechanism in rather more cases than would one with a relatively small interstitial space.

Millea calculated that the interstitial space in silicon is larger than that in germanium. He uses the fact to account for the experimental finding that diffusion of some group III acceptors in silicon can be explained by the mechanism, but diffusion of group III elements in germanium cannot. Experiments on gold diffusion in germanium, however, indicate that the interchange mechanism might well operate. This is reasonable, since gold is a small atom and therefore requires less interstitial space.

6.6 Diffusion in the presence of ion pairing

In Chapter 4 the effect of the coulombic attraction of ionised donors and acceptors on solubility was considered. Pairs forming between one of each type of ion caused a considerable increase in solubility. It might be expected that the formation of these pairs would have an effect on diffusion coefficient and this is so: the phenomenon has been described by Reiss, Fuller and Morin (1956).* The following is essentially the treatment given in their original article.

Consider a homogeneous semiconductor containing a concentration of relatively immobile acceptors. If a mobile donor (e.g. lithium in silicon) is allowed to diffuse into the crystal, ion pairs form, and the diffusion coefficient of the donor is decreased because of the 'drag' effect of the pairing with acceptors. If the concentration of pairs is $\mathscr{P}$ and the concentrations of donors and acceptors are $[D]$ and $[A]$, respectively, then the concentrations of unpaired donors and acceptors are $([D] - \mathscr{P})$ and $([A] - \mathscr{P})$, respectively. A mass-action relation can be written

$$\frac{\mathscr{P}}{([A] - \mathscr{P})([D] - \mathscr{P})} = \Omega \tag{6.32}$$

as before, where Ω is the equilibrium constant and it is assumed that all donors and acceptors not involved in pairs are ionised. Since the crystal is originally homogeneously doped with acceptors, $[A]$ is a constant. The concentration of the diffusing species is not constant, of course, so both $[D]$ and $\mathscr{P}$ are functions of distance in the diffusion direction x. The density of diffusable ions at any point is $([D] - \mathscr{P})$ and it is the gradient of this quantity that provides the driving force for diffusion. Fick's law therefore becomes

$$J = -D_0 \frac{\partial([D] - \mathscr{P})}{\partial x} \tag{6.33}$$

It is convenient to write eqn. 6.32 in a slightly different form

$$\Omega = \frac{-([D] - \mathscr{P}) + [D]}{\{([A] - [D]) + ([D] - \mathscr{P})\}\{[D] - \mathscr{P}\}} \tag{6.34}$$

* Reiss, H., Fuller, C. S., and Morin, F. J., *Bell System Tech. J.*, 1956, **35**, p. 535

This allows an expression to be obtained for $([D] - \mathscr{P})$ as a function of $[D]$.

$$[D] - \mathscr{P} = \frac{1}{2}\left([D] - [A] - \frac{1}{\Omega}\right) + \left\{\frac{1}{4}\left([D] - [A] - \frac{1}{\Omega}\right)^2 + \frac{[D]}{\Omega}\right\}^{1/2} \qquad (6.35)$$

The expression can be substituted into eqn. 6.33 to give

$$J = -\frac{D_0}{2}\left\{1 + \frac{\frac{1}{2}\left([D] - [A] + \frac{1}{\Omega}\right)}{\left\{\frac{1}{4}\left([D] - [A] - \frac{1}{\Omega}\right)^2 + \frac{[D]}{\Omega}\right\}^{1/2}}\right\}\frac{\partial [D]}{\partial x} \qquad (6.36)$$

In the absence of pairing, the diffusion coefficient for the donor would be simply D_0. The effect of the pairing phenomenon is to replace the constant D_0 with a diffusion coefficient D which varies with x, i.e.

$$D = \frac{D_0}{2}\left\{1 + \frac{\frac{1}{2}\left([D] - [A] + \frac{1}{\Omega}\right)}{\left\{\frac{1}{4}\left([D] - [A] - \frac{1}{\Omega}\right)^2 + \frac{1}{\Omega}[D]\right\}^{1/2}}\right\} \qquad (6.37)$$

The expression can be made considerably simpler for the two interesting limiting cases $[D] \ll [A]$ and $[D] = [A]$. Consider first a low concentration of donors, i.e. $[D] \to 0$ Eqn. 6.37 becomes

$$D = \frac{D_0}{1 + \Omega[A]} \qquad (6.38)$$

Now the denominator of eqn. 6.38 can be much larger than unity. A large reduction in diffusion coefficient is therefore predicted for low donor concentrations.

The other extreme, of $[D] = [A]$ is also interesting. Eqn. 6.34 becomes

$$D = \frac{D_0}{2}\left\{1 + \frac{1}{(1 + 4\Omega[A])^{1/2}}\right\} \qquad (6.39)$$

What this means physically can be seen more clearly by considering eqn. 6.32 under the condition $[D] = [A]$. It becomes

$$\Omega = \frac{\mathscr{P}}{([A] - \mathscr{P})^2}$$

Now we are primarily interested here in the situation in which the pairing is strong. In this case $[A] \to \mathscr{P}$ and Ω becomes very large. The second term in the bracket of eqn. 6.39 then goes to zero, and the simple expression is found

$$D = \tfrac{1}{2}D_0 \qquad (6.40)$$

The effect of pairing on donor diffusion when the crystal is homogeneously doped with an acceptor can therefore be summed up as follows. The diffusion coefficient at the leading edge of the diffusion profile, at which $[D] \ll [A]$, is severely reduced. The effect becomes smaller as the crystal surface is approached and $[D]$ becomes larger so that if $[D] = [A]$, only a factor of two is involved.

6.7 Diffusion by divacancies

The difficulties encountered by a substitutional impurity in a compound wishing to diffuse through the crystal have already been noted at the beginning of the chapter. The trouble is that the nearest equivalent site is not a nearest neighbour but a next-nearest neighbour. This means the diffusion jump is rather large, and that it usually involves squeezing between two other atoms. One way this problem can be alleviated is for the jump process to involve divacancies rather than single vacancies. In essence this has the effect of creating a little more empty space in the crystal for the diffusing atom to move through. In the compound *MX* there are three types of divacancy possible, namely $V_M V_X$, $V_M V_M$ and $V_X V_X$, where the last two are really of the same type and do not involve adjacent sites.

Let us start with the first type of divacancy and take as a concrete example a system for which this type of process has been postulated, namely self-diffusion in InSb. The defect under discussion is therefore $V_{In} V_{Sb}$. A simple substitutional process in which an indium atom jumps from an indium site to a vacancy on the next indium site depends on the concentration of indium vacancies. This concentration can be manipulated by carrying out the diffusion experiment several times using different ambient indium vapour pressures. The experiment was carried out by Kendall and Huggins (1969),* who found that the self-diffusion coefficients of both elements were independent of the vapour pressures of indium and antimony and therefore, presumably, independent of the vacancy concentrations. The defect involved in the diffusion process must also be independent of these parameters and it is fairly easy to show that the defect concentration $[V_{In} V_{Sb}]$ obeys this requirement. The mass-action formulation is

$$V_{In} + V_{Sb} \rightleftharpoons V_{In} V_{Sb}$$

$$[V_{In}][V_{Sb}] = K[V_{In} V_{Sb}] \tag{6.41}$$

Now it has already been shown that the product of the single vacancy concentrations is a constant at constant temperature, so it follows from eqn. 6.41 that the divacancy concentration is also a constant. Hence a constant diffusion coefficient would be predicted, in agreement with experiment. It is perhaps worth pointing out at this stage that this is not the only mechanism that gives a diffusion constant independent of external vapour pressures. Another mechanism giving rise to such an effect has already been mentioned in Section 6.4.

* Kendall, D. L., and Huggins, R. A., *J. Appl. Phys.*, 1969, **40**, p. 2750

In the basic jump process an indium tracer atom, for example, would jump to the indium end of the divacancy. This does not destroy the divacancy, it merely changes its position. Now it might be pointed out that this basic jump is exactly what happens in a simple substitutional process, so how does the divancy help? The answer is that the extra antimony vacancy makes the lattice more open at that point, so the jump can take place with much less distortion of the lattice. A more serious objection, intuitively speaking, would seem to be that the mechanism seems to require a rather high concentration of divacancies to give a reasonable diffusion coefficient. It is true, of course, that if the indium and antimony vacancies were charged they might well carry charges of opposite sign and therefore attract each other quite strongly. Kendall and Huggins (1969) did, in fact, carry out a calculation in which the $[V_{In}V_{Sb}]$ concentration was estimated, and they came to the conclusion that at high temperatures the divacancy concentration might even exceed that of single vacancies. The diffusion coefficients measured by them were, nevertheless, very low, varying between about 10^{-16} cm^2 s^{-1} and 10^{-14} cm^2 s^{-1} in the temperature range 475°C to 517°C (InSb melts at 535°C).

Let us now consider the $V_M V_M$ type of divacancy, again by taking as an example a system for which the process has been suggested. Young and Pearson (1970)* have studied the diffusion of sulphur in GaAs. Sulphur is a donor in this semiconductor, sitting on the arsenic site. It was proposed that diffusion took place through the divacancy $V_{Ga}V_{Ga}$. The jump mechanism postulated does not involve a sulphur/arsenic vacancy jump, but a direct exchange between a sulphur atom on an arsenic site and an arsenic atom at the second nearest-neighbour position. This type of process is usually discounted because there is simply not enough room for the two atoms to squeeze past each other: in the GaAs lattice there are two gallium atoms in the way!

The point is made more clearly in Fig. 6.3 which shows the [111] plane of GaAs. The plane is 'corrugated' in the sense that if the arsenic atoms are in the plane of the page, the gallium atoms are a little below. Each element makes a separate close packed plane. If an impurity at site 1 wishes to exchange with an arsenic atom on site 2, it has to push through the space between gallium atoms at 3 and 4. If we conveniently remove these two by forming $V_{Ga}V_{Ga}$, the process becomes feasible, and this is the mechanism suggested.

The diffusion coefficient will depend on the concentration of the mobile complex $V_{Ga}S^+V_{Ga}$. To simplify the notation, let us call this complex $\mathscr{C}_1$. The sulphur can also exist in the more usual form as a donor on an arsenic site, i.e. as $S_{As}{}^+$. Call this $\mathscr{C}_2$. It is assumed that the diffusion coefficient of $\mathscr{C}_1$; D_1 is much larger than that of $\mathscr{C}_2$ which is to be considered immobile. The equilibrium between the mobile and immobile species is given by

$$\mathscr{C}_2 + V_{Ga}V_{Ga} \rightleftharpoons \mathscr{C}_1 \qquad (6.42)$$

* Young, A. B., and Pearson, G. L., *J. Phys. & Chem. Solids,* 1970, **31,** p. 517

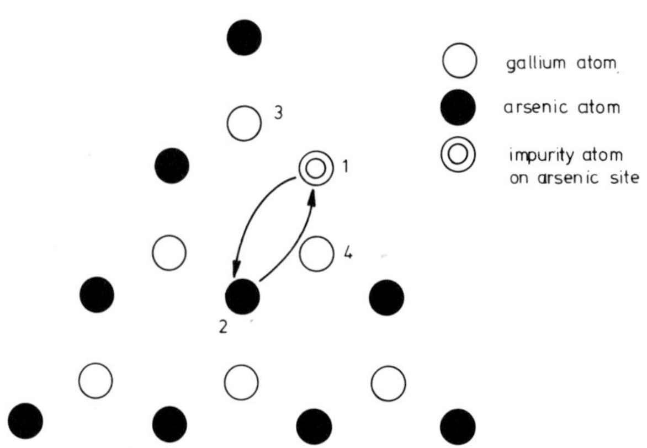

Fig. 6.3 (111) plane of GaAs

leading to the mass-action equation

$$[\mathscr{C}_2][V_{Ga}V_{Ga}] = K_1[\mathscr{C}_1] \tag{6.43}$$

A consideration of the equilibrium between the external gallium vapour pressure and the divacancy gives

$$[V_{Ga}V_{Ga}]P_{Ga}^2 = K_2$$

and since

$$P_{As_4}P_{Ga}^4 = K_3$$

we have

$$[V_{Ga}V_{Ga}] = K_4P_{As_4}^{1/2} \tag{6.44}$$

and since the sulphur exists either in the form $\mathscr{C}_1$ or the form $\mathscr{C}_2$

$$[\mathscr{C}_1] + [\mathscr{C}_2] = [S] \tag{6.45}$$

Solving eqns. 6.43 to 6.45 gives

$$[\mathscr{C}_1] = \left(\frac{K_5P_{As_4}^{1/2}}{1 + K_5P_{As_4}^{1/2}}\right)[S] \tag{6.46}$$

where

$$K_5 = \frac{K_4}{K_1}$$

Since only the $\mathscr{C}_1$ species is deemed to contribute to the diffusion, the effective coefficient is given by eqn. 6.13 as

$$D = D_1 \frac{\partial [\mathscr{C}_1]}{\partial [S]} = D_1 \left(\frac{K_5 P_{As_4}^{1/2}}{1 + K_5 P_{As_4}^{1/2}} \right) \tag{6.47}$$

The variation of diffusion coefficient with arsenic vapour pressure predicted by this model is as follows. At low arsenic pressures, D is low. As the pressure increases, $[\mathscr{C}_1]$ gets larger and the effective diffusion coefficient also increases. Finally, at some high arsenic pressure, almost all of the sulphur is in the mobile form and the diffusion coefficient saturates at the value D_1, which is simply the diffusion coefficient of the complex $\mathscr{C}_1$. All of this behaviour is followed by the experimental results for sulphur in GaAs, which are shown in Fig. 6.4.

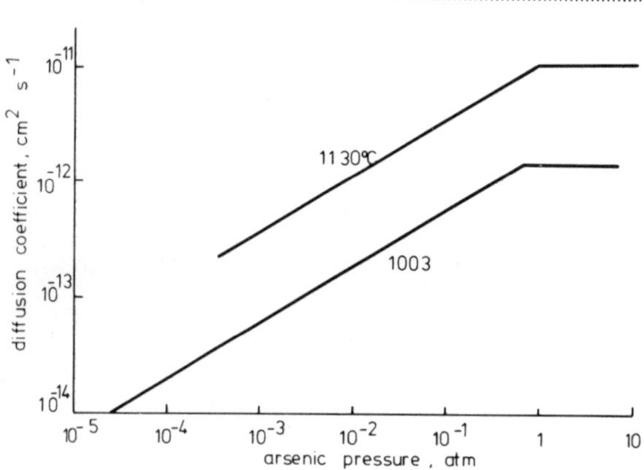

Fig. 6.4 Variation of diffusion coefficient of sulphur in GaAs as a function of ambient arsenic pressure, determined by experiment

Reproduced from Young, A. B., and Pearson, G. L.: *J. Phys. & Chem. Solids*, 1970, **31**, p. 517

6.8 Substitutional-interstitial mechanism

This mechanism is believed to occur in a number of technologically important semiconductor systems and for that reason it is one of the most extensively investigated. It is one of the more interesting examples of two modes of diffusion operating together.

The mechanism was originally suggested by Frank and Turnbull (1956)* to

* Frank, F. C., and Turnbull, D., *Phys. Rev.*, 1956, **104**, p. 617

explain the behaviour of copper diffusing into germanium. It had been found (Tweet and Gallagher, 1956)* that for a given set of diffusion conditions, the depth of penetration of the copper depended very much on the disolocation content of the germanium: the greater the degree of imperfection, the greater the penetration. In crystals containing a reasonably large number of disolocations, it was found that the distribution of copper corresponded to Fick's law close to the diffusion surface, but that at greater depths there was a level of copper concentration which rose with time and was relatively independent of distance. The most obvious interpretation of the observations is that the copper diffuses preferentially along dislocation lines, which provide 'pipes' of easy diffusion. Frank and Turnbull (1956) looked at this idea carefully and showed that it asked rather a lot of the dislocations to achieve the observed effects. The enhancement of diffusion coefficient at a dislocation would have to be unrealistically large: alternatively the diameter of the pipe of easy diffusion would have to be much greater than the usually accepted width of a dislocation line.

Frank and Turnbull (1956) suggested the following alternative explanation. Copper dissolves in germanium in two states; substitutional and interstitial. In the interstitial state it has a high diffusion coefficient and in its substitutional condition, diffusivity is very low. Most of the copper atoms in the crystal are substitutional, however. Conversation from the interstitial to the substitutional state requires a supply of vacancies which can come either from the free surface, for those parts of the crystal close to such a surface, or from dislocation climb for the bulk of the crystal. The process therefore operates more efficiently in a crystal with many dislocations. The implications of this model were worked out by Sturge (1957)† and it is his treatment that will be followed here.

The square-bracket notation becomes rather cumbersome in the following derivation, and it is more convenient to let the solubilities of the substitutional and interstitial species be C_s' and C_i' respectively. These concentrations will be achieved at the surface of the specimen as soon as the diffusion process starts. Within the bulk of the specimen, however, the concentrations will be less than these values and will be functions of penetration x. Call these concentrations C_s and C_i where it is understood that they both depend on x. The two forms of the diffusing element do not move through the crystal independently: they can react as follows:

$$S \rightleftharpoons I + V \tag{6.48}$$

where V is a vacancy. A typical sequence for the process might be described as follows: the atom joins the lattice at the surface from some external phase such as a vapour, it then leaves its substitutional site and goes interstitial, i.e. eqn. 6.48 goes to the right. In the interstitial state it diffuses through the lattice very fast until it finds a vacancy. It then rejoins the lattice by eqn. 6.48 operating to the left.

* Tweet, A. G., and Gallagher, C. J., *Phys. Rev.*, 1956, **103**, p. 828

† Sturge, M. D., *Proc. Phys. Soc. (London)*, 1957, **73**, p. 297

Compared to this process, straightforward substitutional diffusion is so slow that it can be ignored. It is important to realise, however, that although the process has been efficient in moving an atom from the surface to an interior site, it has in fact, used up an interior vacancy. It is possible, therefore, for the concentration of vacancies in the interior C_v to become depressed from the thermal equilibrium value C_v'. In general, therefore, C_v is also a function of x, but at the surface $C_v = C_v'$, because the surface is, in effect, an infinite source of vacancies.

We now make the assumption that the reaction given by eqn. 6.48 takes place so rapidly that at any point in the crystal the interstitials and substitutionals are in instantaneous equilibrium. Under these circumstances, the law of mass action can be applied

$$C_s = KC_iC_v \tag{6.49}$$

The problem has been simplified somewhat by assuming in eqns. 6.48 and 6.49 that the species S, I do not carry a charge. This restriction will be removed in the next section but, just for the moment, the problem is quite complicated enough. For the special case of $x = 0$, i.e. at the surface, we have concentrations C_s', C_i', C_v' giving an expression for K

$$K = \frac{C_s'}{C_i'C_v'} \tag{6.50}$$

If the diffusion coefficient of the substitutional atom is considered negligible, all of it is transferred by interstitial diffusion, and

$$\frac{\partial C_i}{\partial t} + \frac{\partial C_s}{\partial t} = D_i \frac{\partial^2 C_i}{\partial x^2} \tag{6.51}$$

D_i is the interstitial diffusion coefficient. Now consider the continuity equation for the vacancies

$$\frac{\partial C_v}{\partial t} = D_v \frac{\partial^2 C_v}{\partial x^2} - \frac{\partial C_s}{\partial t} + k'(C_v' - C_v) \tag{6.52}$$

D_v is the diffusion coefficient for vacancies. In eqn. 6.52 the first term on the right-hand side represents vacancies gained by diffusion from the surface, the second represents loss of vacancies due to interstitials going substitutional, and the third term is the bulk production of vacancies by dislocation climb and any other process the crystal might be able to manage. The simplest possible assumption is made for this last term, namely that vacancies are produced at a rate proportional to the deviation of C_v from its equilibrium value, with constant of proportionality k'.

The distribution of the diffusing atoms with respect to time and distance should be found by solving eqns. 6.49 to 6.52. This cannot be done explicitly, however, and it is anyway much more useful from the physical point of view to consider a number of important special cases.

(*a*) Vacancy equilibrium is maintained. The production of vacancies in the bulk is then so efficient that the crystal always has the correct number, i.e. $C_v = C_v'$ for all x. Eqn. 6.49 becomes

$$C_i = \frac{C_s}{KC_v'}$$

which, substituted in eqn. 6.51 leads to the diffusion equation

$$\frac{\partial C_s}{\partial t}(1 + KC_v') = D_i \frac{\partial^2 C_s}{\partial x^2} \tag{6.53}$$

Now is already been assumed that only a small fraction of the diffusing element is in the interstitial form at any one time. It follows that C_s is approximately equal to the total impurity concentration and eqn. 6.53 can be taken as the diffusion equation for the diffusing atoms.

Rearranging eqn. 6.53 and substituting for K from eqn. 6.50, we have

$$\frac{\partial C_s}{\partial t} = D_i \left(\frac{C_i'}{C_i' + C_s'}\right) \frac{\partial^2 C_s}{\partial x^2}$$

so the effective diffusion coefficient is

$$D_1 = D_i \frac{C_i'}{C_i' + C_s'} \tag{6.54}$$

(*b*) No production of vacancies in the bulk, i.e. $k' = 0$. This is the other extreme. Here the vacancies required in the reaction given by eqn. 6.48 have to diffuse in from the surface. To put it another way, the mechanism becomes one in which a substitutional atom dissociates into an interstitial and a vacancy, which diffuse independently. Since vacancy diffusion is a slow process compared to interstitial diffusion, the interstitial concentration can be put equal to its equilibrium value C_i' for all x. Substituting this time in eqn. 6.52 for C_v and putting $k' = 0$ yields

$$\frac{\partial C_s}{\partial t}(1 + KC_i) = D_v \frac{\partial^2 C_s}{\partial x^2}$$

which, on substituting for K gives an effective diffusion coefficient

$$D_2 = D_v \frac{C_v'}{C_v' + C_s'} \tag{6.55}$$

So that in this circumstance, the diffusion process is limited by the vacancy diffusion coefficient.

(*c*) $k' > 0$ but still small. We can still make the assumption made for (*b*) that the interstitial process is so much faster than any other that interstitial equilibrium establishes itself almost immediately (relatively speaking) so that we can put $C_i = C_i'$ for all x. Eqns. 6.49 and 6.52 then give

$$\frac{\partial C_s}{\partial t} = D_2 \frac{\partial^2 C_s}{\partial x^2} + \frac{C_s' - C_s}{\theta} \tag{6.56}$$

where $\theta = (C_v' + C_s')/k'C_v'$. Note that θ has the dimensions of k'^{-1}, i.e. of time. Consider a semi-infinite crystal. Eqn. 6.56 must be solved subject to the usual boundary condition:

$$C_s = C_s' \text{ at } x = 0 \text{ for all time.}$$

This can be done by applying the method of Laplace transforms described in Chapter 2. The solution is

$$\frac{C_s}{C_s'} = 1 - \exp\left(-\frac{t}{\theta}\right) \operatorname{erf} \frac{x}{2(D_2 t)^{1/2}} \tag{6.57}$$

The solution is shown plotted in Fig. 6.5 for three different times of diffusion, namely $t/\theta = ½, 1, 2$. For convenience of plotting, θ has been taken as the unit

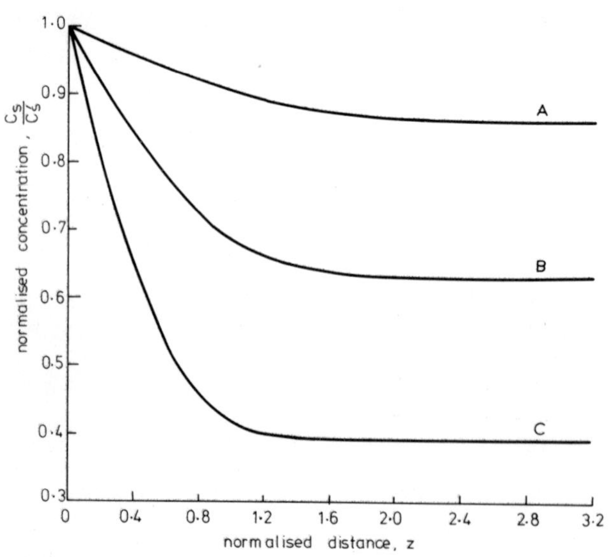

Fig. 6.5 Theoretical profiles determined from eqn. 6.57. The curves A, B, C, are plotted for values of t/θ of 2, 1, 0·5 respectively

of time and instead of x, a 'normalised' distance given by $z = x/2(D_2\theta)^{1/2}$ has been used. Close to the surface the erf term in eqn. 6.57 varies quite sharply with distance and the profile drops away roughly as an error function complement solution characterised by the diffusion coefficient D_2. Physically this indicates that in the region close to the surface, the surface itself is the principal source of vacancies. Further into the specimen, however, the surface is too remote to be a useful source and the production of vacancies by dislocation climb becomes the main means of support. For eqn. 6.57 this is manifested by the erf term going to unity as x (and z) get sufficiently large. The equation then becomes

$$\frac{C_s}{C_s'} = 1 - \exp\left(-\frac{t}{\theta}\right) \tag{6.58}$$

The distribution is then independent of distance, but the constant value increases with time towards C_s'. What this means in physical terms is that interstitial atoms are going substitutional as vacancies become available from the bulk generation process.

The solution has a rather interesting consequence, when applied to a real experiment. Real crystals are not semi-infinite, but providing they are not very thin, the above solution will still apply under the appropriate experimental conditions. Consider diffusing into a thick slice with surfaces at $x = 0$ and $x = d$ from a vapour phase at $x < 0$. According to our present assumptions the interstitial concentration C_i' establishes itself almost immediately and a substitutional concentration, greater than C_i' builds up from this concentration by way of the reaction given by 6.48 going to the left. The large buildup close to $x = 0$ occurs because the surface is a source of vacancies. It is also true, however, that the surface at $x = d$ is a large vacancy source and, according to the above argument, a substitutional concentration greater than that in the bulk should build up close to the back surface of the specimen. This would give rise to a U shaped diffusion profile, showing apparent 'uphill' diffusion close to the back surface. Profiles of this type have often been observed. Fig. 6.6 shows a radio-tracer profile for gold diffusing in silicon. The gold was originally plated on to one side of the silicon slice but still shows the enhanced concentration at both surfaces.

This enhancement at the back surface forms the basis of a rather elegant experimental test for the substitutional-interstitial mechanism. The method is due to Kendall (1968),* and he calls it a 'figure test'. A radio tracer of the diffusant is deposited in the form of a geometric figure on the front of a slice, and it is taken up to diffusion temperature for an appropriate time. Autoradiographs are then taken of the front and back surfaces, and also of the

* Kendall, D. L., 'Semiconductors and Semimetals' Vol. 4, Ed. Willardson and Beer, (Academic Press, 1968)

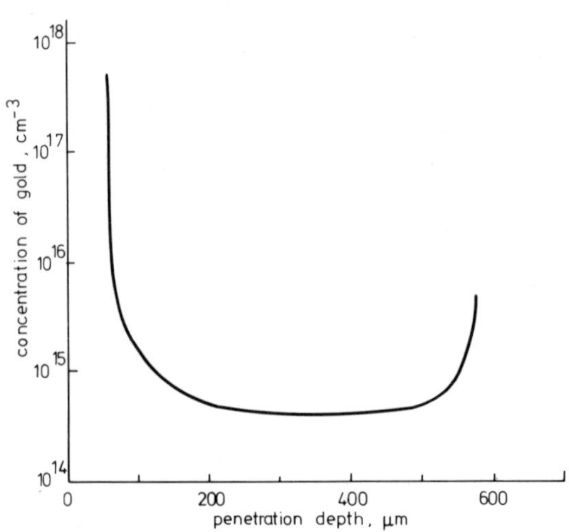

Fig. 6.6 Radiotracer profile for gold diffusing in silicon at 1000° C, diffusion time 30 min.

Reproduced from Spokel, G. J., and Fairfield, J. M.: *J. Electrochem. Soc.,* 1965, **112**, p. 200

middle, after lapping away half the specimen. If the substitutional-interstitial mechanism is operative with a small k' value, then the traces of the geometric figure seen on the front and back surfaces are much stronger than any trace in the middle.

(*d*) This is for large k' (small θ). The crystal can now provide bulk vacancies quite efficiently to help the interstitial-substitutional transition. It is now no longer valid to assume that the interstitial concentration assumes its equilibrium value C_i' before anything else starts to happen, and the vacancy concentration C_v, although still less than the thermal equilibrium value C_v', is only slightly depressed. We assume that the dislocation mechanism is so effective at replacing vacancies that diffusion from the surface can be ignored. A quasi-equilibrium can then be reached in which interstitial atoms can join the lattice only at the rate at which the mechanism creates vacancies for them, i.e.

$$\frac{\partial C_s}{\partial t} = k'(C_v' - C_v) \tag{6.59}$$

Now consider the interstitial concentration. If k' were so large that the vacancy concentration was always at its equilibrium value C_v', we would have for the semi-infinite specimen

$$\frac{C_i}{C_i'} = \frac{C_s}{C_s'} = \operatorname{erfc} \mu$$

where $\mu = x/2(D_1 t)^{1/2}$. If k' is not quite so large, it can be shown that to a first approximation C_i/C_i' is unchanged. Physically, this amounts to saying that a lack of vacancies will interrupt the flow of atoms from the interstitial distribution to the substitutional distribution, as described by eqn. 6.48, but will not interfere very much with the original interstitial distribution, providing the deviation from equilibrium is not very great.

Substituting $C_i = C_i' \operatorname{erfc} \mu$ into eqn. 6.49 and combining with eqn. 6.59 gives

$$\frac{\partial C_s}{\partial t} = \frac{k'}{KC_i'}\left(C_s' - \frac{C_s}{\operatorname{erfc} \mu}\right)$$

Using eqn. 6.50 to remove the equilibrium constant and approximating θ by $C_s'/k'C_v'$ gives

$$\theta \frac{\partial C_s}{\partial t} = C_s' - \frac{C_s}{\operatorname{erfc} \mu} \tag{6.60}$$

We have now simplified the problem quite considerably, but further simplification is required to solve eqn. 6.60. The approximate solution is given by Sturge as

$$C_s = C_s' \operatorname{erfc} \mu \left[1 - \exp\left(\frac{-t}{\theta \operatorname{erfc} \mu}\right)\right] \tag{6.61}$$

in which μ has been treated as if it were a constant. The error involved in the approximation can be shown to be small except in the region $t/\theta \simeq 10 \operatorname{erfc} \mu$.

Eqn. 6.61 must obviously be used with some caution since so many approximations were used in its derivation. However, it is worth noting that as erfc $\mu \to 1$, eqn. 6.61 approaches eqn. 6.58. The equation can also be interpreted in simple physical terms. The first term on the right-hand side represents the 'equilibrium' solution to the problem, i.e. the form of the profile obtained when vacancy equilibrium is maintained and $C_v = C_v'$ throughout the experiment. This term is modified by the expression in brackets, which is always less than unity. The distribution is therefore always less than the equilibrium distribution. As time progresses, the exponential term gets smaller and the nonequilibrium solution approaches the equilibrium solution ever more closely. This behaviour is indicated in Fig. 6.7. The dotted curve D is the equilibrium profile given by

$$C_s = C_s' \operatorname{erfc} \mu \tag{6.62}$$

It is plotted as a function of $x/t^{1/2}$ so that as presented it does not vary with time. The curves A, B, C, are given by eqn. 6.61 for different times with $t_A < t_B < t_C$. To sum up the model, the 'equilibrium' interstitial distribution occurs immediately the diffusion process starts, but the equilibrium substitutional distribution does not, because of a lack of vacancies. As time passes, vacancies are produced by dislocation

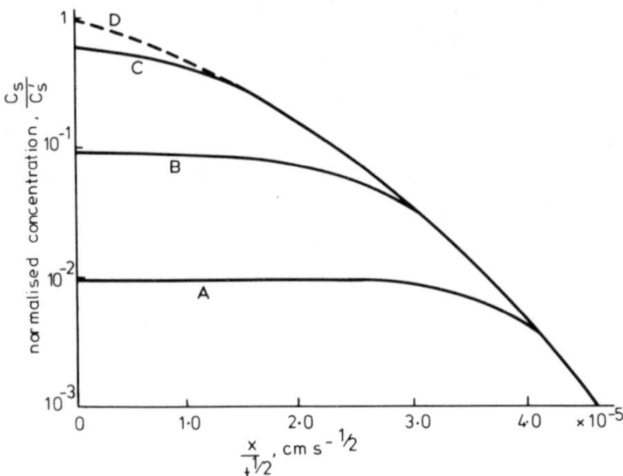

Fig. 6.7 Theoretical profiles determined using eqn. 6.61, with a value of 10^{-10} cm^2 s^{-1} for D_1. Curve D is an error function complement. Curves A, B, C were plotted using values of t/θ of 0·01, 0·1, 1 respectively

climb and eqn. 6.48 operates at a rate determined by the rate of production of vacancies.

Some thought is required before comparing eqn. 6.61 to experimental results. The striking characteristic about Fig. 6.7 is that the surface concentration predicted is not C_s', but some smaller value C_0, say. C_0 approaches C_s' as $t \to \infty$: call it the 'quasi-surface concentration'. Now this behaviour is contrary to our usual assumptions about surface concentration: we usually assume that the surface assumes the equilibrium solubility at the start of the experiment. This is expressed in mathematical terms by the boundary condition

$$C_s = C_s' \text{ at } x = 0 \text{ for } t > 0$$

and it should be noted that we did not solve eqn. 6.60 subject to this condition. This seems intuitively unreasonable, since the assumption of thermal equilibrium at the surface should not be dependent on the mechanism for producing vacancies in the bulk. The paradox can be resolved as follows. In setting up this particular model, we decided to ignore the effect of any vacancies coming in from the surface. This is quite satisfactory, physically speaking, if k' is large, except in the region very close to the surface. In the extreme case of $x = 0$, the surface is an infinite source of vacancies and it is certainly not true to say that the supply at this point comes from dislocation climb. At $x = 0$ and a thin region just below the surface, the equilibrium concentration of vacancies, C_v' is maintained by the surface. For this region the distribution is given more accurately by eqn. 6.62 than by eqn. 6.61. However, in the bulk of the material, the model is valid and eqn. 6.61 is followed.

This line of argument is interesting, because it predicts a very peculiar shape for a diffusion profile. The distribution starts at $x = 0$ with concentration C_s' and approximately follows curve D (Fig. 6.7) for a short distance into the material. It then drops down to a curve such as A, B or C and follows it for the rest of the way. This gives rise to a curve with a concave section. Such curves are often observed in semiconductor diffusion work, especially with III–V compounds and the point will be taken further in the next Section. The idea is used in Fig. 6.8 to fit some of the results of van der Maeson and Brenkman (1955)* for diffusion of copper in germanium at 800°C. The points are experimental for a diffusion time of 20 min. The upper curve follows eqn. 6.62 and the lower one is eqn. 6.61, with $t/\theta = 9 \times 10^{-2}$. Note that the quasi-surface concentration is more than an order of magnitude lower than the true surface concentration.

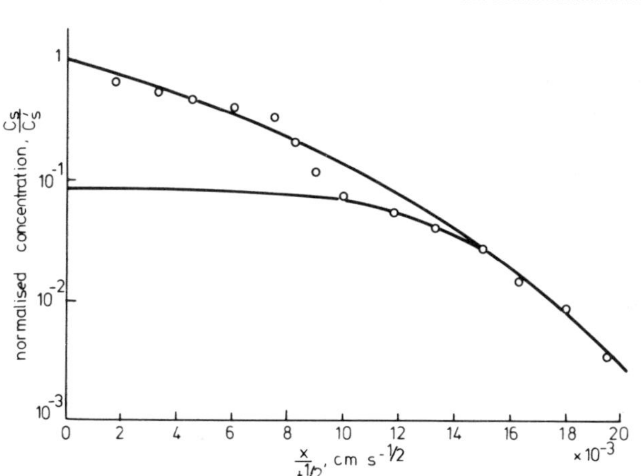

Fig. 6.8 Experimental results of van der Maesen and Brenkman fitted to the theory of Section 6·8, using $D_1 = 2{\cdot}3 \times 10^{-5}$ cm^2 s^{-1} and $t/\theta = 0{\cdot}09$. The results are for a 20 minute diffusion of copper in germanium at 800° C

6.9 Substitutional-interstitial mechanism with charged diffusing species

In the previous Section it was assumed that the diffusing atoms are uncharged. In general this would not be so: the atoms would be charged donors or acceptors. In this Section we deal with the case of charged species and to fix ideas more easily take the specific case of zinc diffusing in GaAs. This is one of the most extensively

* van der Maeson, F., and Brenkman, J. A., *Trans. Electrochem. Soc.*, 1955, **102**, p. 229

investigated of semiconductor diffusion systems, largely because of its technological importance. The details given in this Section, together with those on the solubility of zinc in GaAs given in Section 3.17 provide a resumé of the knowledge to date concerning this diffusion system. It also gives some idea of how much information has to be gathered before a good understanding of such a system can be achieved.

Zinc is a group II element which occupies the gallium site of GaAs, becoming the ionised substitutional acceptor Zn_s^-. Once again, we assume that a very small fraction of the zinc is interstitial, and also assume that in this state it loses an electron to become the ionised interstitial donor Zn_i^+. The diffusion coefficient of the substitutional species is negligibly small. An equation similar to eqn. 6.48 relates the two zinc species

$$Zn_i^+ + V_{Ga} \rightleftharpoons Zn_s^- + 2h$$

where h is a hole and has to be introduced to maintain charge neutrality. The law of mass action gives

$$C_i^+ C_v = KC_s^- (\gamma p)^2$$

and it should be noted that C_v refers to the concentration of gallium vacancies. If the semiconductor is not degenerate, it can be assumed to a good degree of accuracy that all atoms are charged, and the positive and negative signs can be dispensed with, i.e. $C_i^+ = C_i$, $C_s^- = C_s$ and C_s is approximately equal to the total zinc concentration.

Consider first of all a situation in which the zinc doping turns the semiconductor p type i.e. $C_s > n_i$. The condition for electrical neutrality gives $C_s \simeq p$ and since $\gamma = 1$ except for very high concentrations (Chapter 4) the mass-action equation simplifies to

$$C_i C_v = KC_s^3 \tag{6.63}$$

with the special case of equilibrium

$$C_i' C_v' = KC_s'^3 \tag{6.64}$$

Eqns. 6.51 and 6.52 will hold, so once again there are four equations to consider. As in the preceding Section, it is simplest to consider special cases.

(a) For vacancy equilibrium maintained, $C_v = C_v'$. This is most often assumed. Eqn. 6.63 is used to substitute for C_i in eqn. 6.51. When K is eliminated using eqn. 6.64, the expression becomes

$$\frac{\partial C_s}{\partial t}\left\{\frac{3C_s^2 C_i'}{C_s'^3} + 1\right\} = \frac{\partial}{\partial x}\left\{\frac{3C_i' D_i}{C_s'^3} \cdot C_s^2 \cdot \frac{\partial C_s}{\partial x}\right\} \tag{6.65}$$

Now C_s can only be less than C_s' and, by postulate, $C_i' \ll C_s'$, so it follows that

$3C_s^2 C_i'/C_s'^3 \ll 1$. Allowing for this eqn. 6.65 assumes the form of Fick's law with an effective diffusion coefficient

$$D_3 = \frac{3C_i'D_i}{C_s'^3} C_s^2 \tag{6.66}$$

A concentration-dependent diffusion coefficient is obtained, varying as the square of the zinc concentration.

This theory can, in fact, be checked experimentally using a method called the isoconcentration technique. The technique will be mentioned further in the following chapter on experimental methods, but can be summarised as follows. A radio-tracer diffusion is carried out using a specimen of GaAs that, instead of being as pure as possible in the first instance, is already homogeneously doped with zinc to a level C_s'. All that happens during the diffusion is that radioactive zinc diffuses into the semiconductor and nonradioactive zinc diffuses out. The total amount remains constant at C_s'. One might say that as far as the crystal is concerned, nothing is happening. This is not a process that uses up vacancies, so vacancy equilibrium is maintained and $C_v = C_v'$ as required by this theory. Since the crystal remains homogeneously doped throughout, an erfc diffusion profile is obtained, indicating the single value of D_3 corresponding to $C_s = C_s'$.

However, the solubility C_s' of zinc in GaAs is not fixed at constant temperature, but depends on the vapour pressure of zinc outside the GaAs during the diffusion. Thus C_s' can be varied from experiment to experiment by varying this vapour pressure. A series of values of D_3 can therefore be obtained for different values of C_s', and a graph of diffusion coefficient against concentration can be plotted (Kadhim and Tuck, 1972),* and compared to eqn. 6.66. Such a graph is shown in Fig. 6.9: a square-law dependence is indicated to a high degree of accuracy.

A further useful experimental check can be made on eqn. 6.66. Comparison of eqns. 6.64 and 6.66 shows that the latter can be written

$$D_3 = \frac{3KD_i}{C_v'} C_s^2 \tag{6.67}$$

The concentration of vacancies in a compound at thermal equilibrium has already been considered at some length. A quantity such as C_v' is not fixed uniquely by temperature, but also depends on the vapour pressure of arsenic or gallium in the surrounding atmosphere. Since arsenic occurs as As_4 at high temperatures, it works out that $C_v' \propto P_{As}^{1/4}$. A second series of experiments can therefore be carried out at a higher arsenic vapour pressure to check the predicted relationship $D_3 \propto P_{As}^{-1/4}$. This set of results is shown in Fig. 6.10. The same square-law dependence is shown as in Fig. 6.9, but the effect of having a higher ambient arsenic pressure is to remove the line down to lower values of D_3.

* Kadhim, M. A. H., and Tuck, B., *J. Mat. Sci.*, 1972, 7, p. 68

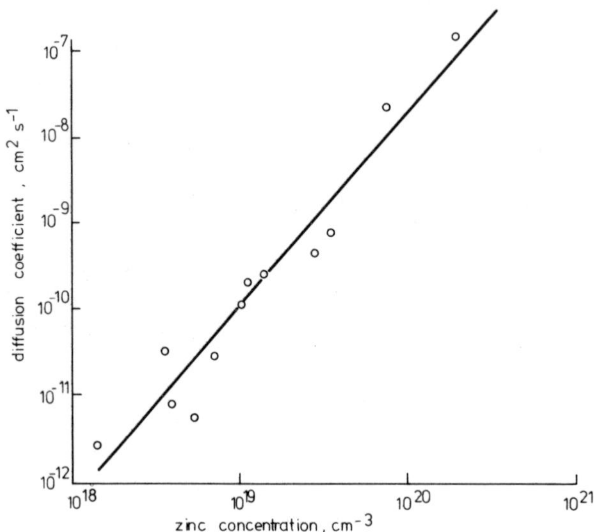

Fig. 6.9 Variation of diffusion coefficient of zinc in GaAs as a function of zinc concentration. The results were taken at 1000°C at dissociation pressure of arsenic (approx. 10^{-4} atm)

Reproduced from Kadhim, M. A. H., and Tuck, B.: *J. Mat. Sci.*, 1972, **7**, p. 68

It is perhaps useful to consider in simple physical terms why a concentration-dependent diffusion coefficient is found. We are dealing with the case for which the zinc doping is always greater than the intrinsic carrier concentration at the diffusion temperature. This means that the diffusing zinc dopes the semiconductor p type. The diffusion coefficient depends on the proportion that is interstitial, since it is the interstitial atoms that do the diffusing, i.e. it depends on the ratio C_i/C_s. Now because these species are respectively donor and acceptor, the ratio depends on the Fermi level position. In turn the Fermi level depends on the concentration of zinc acceptors. Hence the ratio depends on C_s, and so does the diffusion coefficient. Looked at in this way, it is clear why the isoconcentration technique works. The total amount of zinc in the specimen is unchanged throughout the experiment, so the Fermi level is fixed. Hence a single value of D_3 is obtained.

Another situation in which the Fermi level remains fixed throughout the experiment is that for which the zinc doping is less than n_i, where n_i is the intrinsic carrier concentration at the diffusion temperature. The GaAs remains intrinsic and the Fermi level stays at the centre of the forbidden gap. We can still write $C_v = C_v'$ as before, and $\gamma = 1$ but the neutrality condition changes to $p = n_i$. Substituting

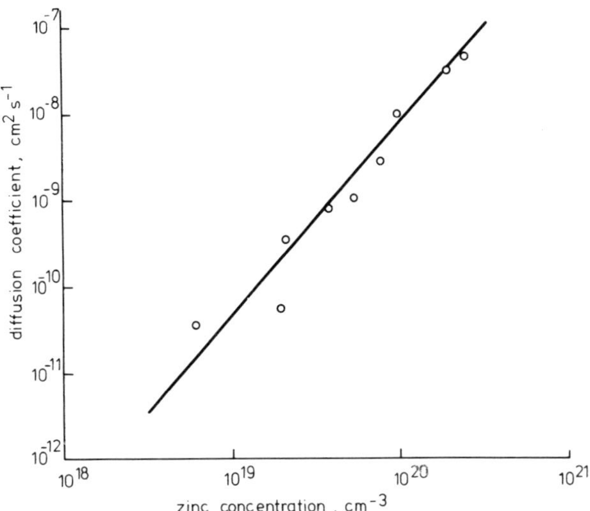

Fig. 6.10 Variation of diffusion coefficient of zinc in GaAs as a function of zinc concentration. The results were taken at 1000°C at an arsenic pressure of 4×10^{-2} atm

Reproduced from Kadhim, M. A. H., and Tuck, B.: *J. Mat. Sci.*, 1972, **7**, p. 68

these values into eqn. 6.51 gives a constant value for diffusion coefficient.

$$D_{\text{intrinsic}} = \frac{Kn_i^2 D_i}{C_v'}$$

(b) When there is no production of vacancies in the bulk, i.e. $k' = 0$, the mechanism for the diffusion of zinc atoms is that the substitutional species dissociates into vacancies and interstitial zinc, which diffuse independently. The high interstitial diffusivity then maintains C_i close to C_i'. A mathematical argument very similar to that given above, substituting this time into eqn. 6.52, gives an effective diffusion coefficient

$$D_4 \simeq \frac{3C_v' D_v}{C_s'^3} \cdot C_s^2$$

As might be expected, the coefficient now depends on that of the gallium vacancies, D_v. It is worth noting, however, that D_4 is also proportional to the square of the zinc concentration.

(c) When there is a finite production rate of vacancies in the bulk, the parameter k'

must be taken into account. In the previous section, two cases were considered, namely that for very small k' and that for k' large enough almost to maintain vacancy equilibrium. It is the latter that is of most interest in the diffusion of zinc in GaAs and this will now be considered.

The line of argument is very similar to that followed in Section 6.8 (d). First consider the situation when vacancy equilibrium is maintained so that $C_v = C_v'$ throughout. It has been shown above that this leads to a diffusion equation of the form

$$\frac{\partial C_s}{\partial t} = \frac{\partial}{\partial x}\left(D_0 C_s^2 \frac{\partial C_s}{\partial x}\right) \tag{6.68}$$

The equation does not lend itself to any simple solution, but it has been solved numerically using a computer, and the solution is published (Weisberg and Blanc, 1963),* so it can be treated as a known function. Let the solution be called

$$C_s = C_s' F(x, t) \tag{6.69}$$

where $F(x, t)$ varies in the range 0 to 1 and is unity at $x = 0$ for all $t > 0$. From eqns. 6.63 and 6.64

$$C_i = C_i' F^3 \tag{6.70}$$

Once again, we have put $\gamma = 1$ and $p = C_s$, i.e. relatively high zinc concentration is assumed.

Now consider the case when C_v is allowed to fall below C_v'. Eqn. 6.69 will no longer hold for substitutional zinc. Let us assume, however, that to a first approximation eqn. 6.70 is still a reasonable description of the interstitial distribution. This is the same assumption that was made in Section 6.8 (d). Then we have

$$C_v = \frac{C_v'}{(C_s' F)^3} C_s^3 \tag{6.71}$$

Eqn. 6.59 will be as good an approximation here as it was previously, so eqn. 6.71 can be substituted in to give

$$\frac{\partial C_s}{\partial t} = k' C_v' \left\{1 - \left(\frac{C_s}{C_s' F}\right)^3\right\} \tag{6.72}$$

This equation has also been solved numerically (Tuck and Kadhim, 1972)† and a series of solutions is shown in Fig. 6.11 for different diffusion times. Qualitatively the behaviour is the same as the function of eqn. 6.61. The solution shows values of concentration which are less than those given by the 'equilibrium' distribution $C_s' F(x, t)$ for all values of x. As time passes the profile builds up towards $C_s' F(x, t)$

* Weisberg, L. R., and Blanc, J., Phys. Rev., 1963, **131**, p. 1548

† Tuck, B., and Kadhim, M. A. H., *J. Mat. Sci.*, 1972, 7, p. 585

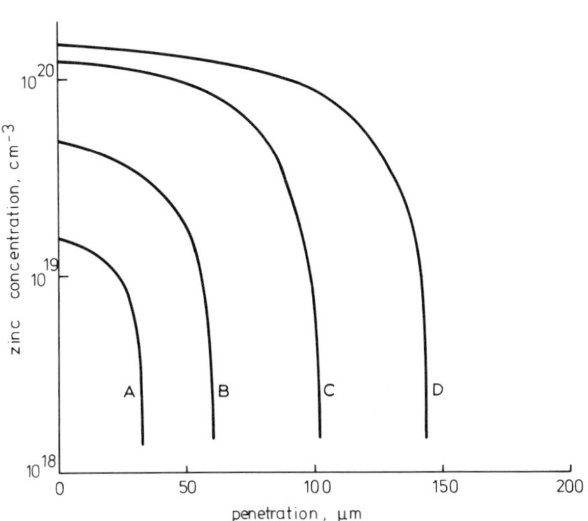

Fig. 6.11 Theoretical 'bulk' profiles of zinc in GaAs with k′ as a constant. Diffusion times are:

Profile A 10 min
Profile B 30
Profile C 90
Profile D 3 h

Reproduced from Tuck, B., and Kadhim, M. A. H.: *J. Mat. Sci.,* 1972, **7**, p. 585

and the quasi-surface concentration C_0 approaches the equilibrium concentration C_s'. Once again thought must be applied before relating these solutions to experiment. The solutions of Fig. 6.11 can only be solutions for the bulk of the crystal, since vacancy equilibrium must always obtain very close to the surface and, in particular, at the surface $C_s = C_s'$ for all time. It seems reasonable to expect, therefore, that at $x = 0$ the diffusion profile should start with the form of eqn. 6.69. In the bulk, however, eqn. 6.72 may be more appropriate and the profile may drop down to one of the curves of Fig. 6.11. This would give a concave region to the profile, in much the same way as was shown in Fig. 6.8. Diffusion profiles of this form are, in fact, quite commonly found in III-V compounds. Fig. 6.12 shows a series of such profiles for zinc diffusing in GaAs at 1000°C. The experimental conditions were identical for all the curves shown, but different diffusion times were used. In the Figure the 'bulk' profiles have been extrapolated back to the surface to obtain estimates for the quasisurface concentration, C_0. It can be seen that C_0 approaches C_s' with increasing time, as predicted above.

In a qualitative way, then, the theory agrees quite well with many experimental results in compound semiconductors. The curves do not agree in detail, however,

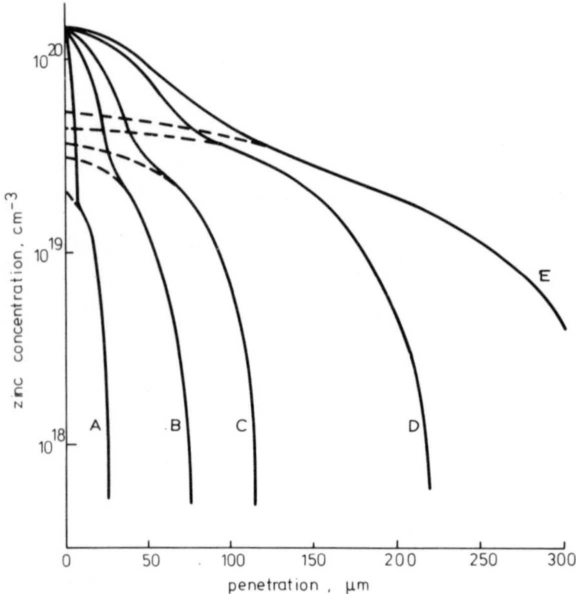

Fig. 6.12 Set of experimentally-determined profiles for zinc in GaAs at 1000°C. Diffusion times were:

Profile A 10 min
Profile B 90
Profile C 3 h
Profile D 9
Profile E 30

Reproduced from Tuck, B., and Kadhim, M. A. H.: *J. Mat. Sci.*, 1972, **7**, p. 585

and it is possible that this is owing to the simplification of taking k' to be a constant. Assuming that production of vacancies in the bulk is primarily owing to the climb of dislocations, k' will be related to the density of dislocations (which may itself be related to zinc concentration) and to the rate at which they can climb. It might therefore, be a complicated function of time and of the concentrations of zinc atoms and gallium vacancies.

Chapter 7

Diffusion techniques

To carry out a diffusion process it is necessary to bring an external phase containing the diffusant into contact with the semiconductor at an elevated temperature. The phase might be a vapour or a liquid or solid. There are a number of ways in which this can be achieved, and this chapter considers some of them. Various ways of plotting diffusion profiles and determining diffusion coefficients from experimental data are then discussed. The chapter is kept general, dealing in the main with principles of experimental technique. References are given to texts dealing with details of the experimental methods.

The most suitable diffusion method depends partly on the diffusion system and partly on the intention of the experiment. If the main purpose of the diffusion is to elicit information on the diffusion mechanism, then it is preferable to have it take place in a sealed ampoule containing known amounts of semiconductor and diffusant. The ampoule then becomes a system about which everything is known and, at least in principle, the theory of Chapter 3 can be used to find out the parameters important to the diffusion process, such as the composition of any condensed phases, and the vapour pressures.

If the intention is to use the diffusion process to prepare quantities of semiconductor devices, however, the sealed-tube method becomes impractical. The business of sealing off an ampoule at the start of every experiment and breaking it at the end is too much trouble and a constant-flow method becomes more useful. The diffusant enters one end of a long tube as a vapour, flows over the semiconductor and leaves at the other end. Another technique, usually called the box method, is something of a compromise between the two. These methods will be considered separately, but it is worth initially looking at some aspects which are common to all three.

7.1.1 Diffusion furnaces

Most diffusion processes employ horizontal cylindrical furnaces. The heat is provided by resistance coils, wound around a furnace tube and made of a high temperature alloy. Such furnaces can reach a temperature of about 1350°C, which is quite high enough for most semiconductor diffusion processes. The one great exception here among common semiconducting materials is SiC, which commonly requires temperatures close to 2000°C for diffusion. An induction furnace has been used for this material, and probably it is the best solution. A resistance furnace using a graphite element has also been used.

The main requirements for a diffusion furnace are; temperature stability over the time occupied by the diffusion, and a reasonably long zone in which the temperature is constant to within ±½°C. Usually a 3-zone furnace is employed, the centre zone being used to carry out the diffusion. The two end sections are there to keep up the temperature at the ends of the central section. The temperatures of the end furnaces are adjusted to give the flattest possible temperature profile within the centre furnace.

There are two main ways whereby temperature control can be maintained, one a little more sophisticated than the other. The simpler method is a single-point system in which one thermocouple is used to measure the temperature at the middle of the centre zone. The voltage generated by the thermocouple is compared to a reference voltage. This second voltage is set to the value the thermocouple would read if it were at the desired temperature. The difference between the two voltages is therefore a measure of the deviation of the furnace from the desired temperature, and can be amplified to drive the control element which provides the current to the furnace windings. In this system all three zones are driven in parallel, so a request for further power from the centre zone will cause more to be delivered to all three. A flat temperature profile in the central region is obtained by adjusting rheostats which are included on each of the three control elements, thereby varying the distribution of power between the three elements.

The system works well under steady-state conditions, but can take a long time to return to the steady state if it is disturbed. At the start of a diffusion run, the boat containing the specimens is introduced into the central zone. Since the boat is cold, the temperature of the central zone is depressed and it asks for more power. Extra power is then delivered to all three zones, even though the two end zones do not need it. The end temperatures increase to rather high values, and only return to the correct ones when the central zone temperature is right. The effect might take several minutes to correct itself, and if the diffusion run is itself only a matter of minutes, it is of some importance. For long diffusion runs the perturbation at the beginning of the experiment is of little importance.

A better arrangement is the slave-control system. The central zone has three thermocouples and each of the end zones has one. One of the central thermocouples is used, just as in the single point system, to maintain the temperature of the centre zone. This is done, as before, by comparison with a set reference voltage. An end-zone thermocouple is connected back-to-back with one of the remaining centre zone thermocouples. The voltage obtained is the difference between the thermocouple emfs, and is a measure of the difference in temperature between the end zone and centre zone. The voltage is compared to a secondary reference voltage and any difference between the two is amplified and used to control the current delivered to the end zone. The mechanism therefore maintains a constant difference in temperature between the end zone and the centre zone. The third central zone thermocouple is used in exactly the same way to control the temperature of the other end zone. The furnace is therefore set up as follows. The primary reference voltage is set to obtain the required temperature in the central zone. The two secondary reference voltages are then set (this can be done independently) to give the temperature differences between the middle and the ends which gives the maximum length of flat temperature profile in the centre of the furnace.

When a cold sample holder is put into the central zone, causing a temperature drop in the centre, a temperature drop also occurs in the end zones. This is because the furnace will always try to keep constant temperature differences between the middle and the ends. The flat temperature profile is therefore not disturbed and the temperature rises up to the correct value uniformly in all three zones.

The furnaces are usually lined with ceramic tubes made from high-purity alumina or mullite. The actual diffusion apparatus, i.e. tubes, diffusion boat, etc. is usually made from silica glass. This material will withstand temperatures up to about 1300°C, although it is subject to devitrification and thermal fatigue and cannot be used indefinitely. If silica is heated for a long time at 1300°C, it can start to deform plastically, just under its own weight. Before the diffusion experiment, the silica-ware must be thoroughly cleaned. It should first be washed in detergent then, after rinsing in deionised water, soaked in aqua regia ($3HCl : 1HNO_3$) for about an hour. It is then rinsed again and dried in an oven.

7.1.2 Preparation of sample

Whatever diffusion technique is used, it is necessary to prepare the semiconductor before putting it into the apparatus. Usually the semiconductor is used in the form of a slice, some hundreds of microns thick. Ideally the surfaces should be flat, shiny, quite clean and damage-free. When the slice is originally received it is none of these things. In general, it will have been cut from a single-crystal ingot with a diamond saw and will have a matt appearance. The cutting will have given rise to regions of crystal damage just below the cut surfaces. The surfaces can be made flat and shiny by mechanical polishing, but this serves to increase the region of damage. Mechanical polishing is sometimes used as a first stage in sample preparation, however, in order to start off with a flat surface. The damaged layer is then removed either by chemical polishing or electropolishing.

Many chemicals dissolve semiconductors, but not many are suitable for the job of polishing, and a good deal of work has been carried out to find suitable solutions. A liquid which is satisfactory for one material is not necessarily suitable for another, so each semiconductor must be treated as a separate case. Most etching solutions are preferential in the sense that they etch different crystal faces at different rates. They usually do not give very flat surfaces but demonstrate interesting surface features which again vary from one crystal orientation to another. Any crystal defects at the surface tend to be shown up, in fact chemical etching is one of the standard ways of determining the dislocation density in a crystal. This is done by counting the density of 'dislocation etch pits' revealed by the chemical polish. However, it is usually possible to find a suitable polishing liquid for a given semiconductor. A simple polishing technique is one in which the slice is immersed in the liquid in a beaker which is rotated about its axis. The axis can also be inclined to the vertical so that the slice rolls around the bottom of the beaker. Liquid passes over the specimen in an even flow and a surface of high quality can be produced. The dissolution process can often be aided by shining a light on the semiconductor. The reason for this is that electrons and holes can take part in the process, and the numbers of carriers are increased if the sample is illuminated.

Chemical polishing of semiconductors is still something of an art rather than a science. An article by Gatos and Lavine (1965)* gives a comprehensive list of chemical etches and polishing liquids for a variety of semiconductors.

Electropolishing is a slightly more complicated procedure but it can produce very good results. The sample to be polished is made the anode of an electrolytic cell and anodic dissolution occurs. Fig. 7.1 is a diagram of an apparatus used by Sullivan, Klein, Finne, Pampliano and Kolb (1963)† to electropolish silicon and germanium. The cathode is a large metal disc, rotated about its centre by a motor. The specimens to be polished comprise the anode, and these are mounted on a smaller disc, situated near the rim of the cathode. The anode disc is also able to rotate (and does so under the action of the differential velocity developed across it by the cathode). The electrolyte is dripped into the very small space between anode and cathode and, because of the fairly high viscosity of the liquids used, a film of electrolyte is always present. If the viscosity is not large enough to maintain the film, then a small amount of glycerine can be added to the liquid. The effect of both discs rotating is to give a strong stirring motion to the liquid, and an even flow of electrolyte over the anode. The very small separation between anode and cathode gives rise to a risk of the electrodes touching. This can be alleviated by using a layer of material such as cotton or paper between the electrodes. The use of such a separator eases the mechanical and operating tolerances.

Holes take part in the anodic dissolution process and this was demonstrated by Sullivan, Klein, Finne, Pampliano and Kolb (1963) in an interesting way. Whereas the apparatus described above gave good results with p-type semiconductor material, in which there is no shortage of holes, n-type semiconductor was not polished satisfactorily. The surface was very uneven, showing deep etch pits. The problem was solved by illuminating the anode surface during polishing, creating

*Gatos, H. C., and Lavine, M. C., Progress in semiconductors, Vol 9, Ed. Gibson, A. F., and Burgess, R. E. (Heywood, 1965)

†Sullivan, M. V., Klein, D. L., Finne, R. M., Pampliano, L. A., and Kolb, G. A., *J. Electrochem. Soc.*, 1963, **110**, p. 412

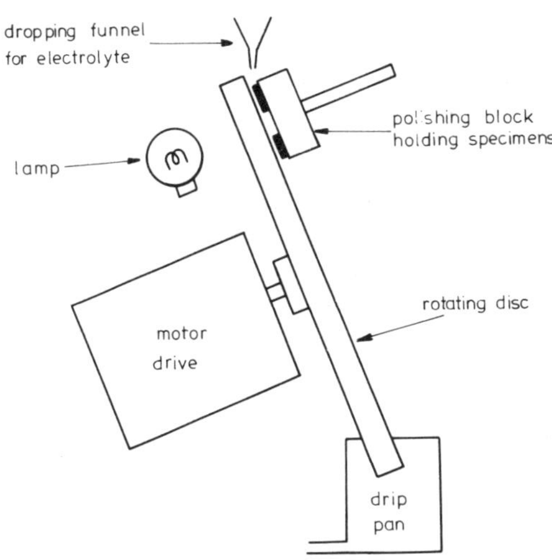

Fig. 7.1 Apparatus for electropolishing silicon and germanium

Reproduced from Sullivan, M. V., Klein, D. L., Finne, R. M., Pampliano, L. A., and Kolb, G. A.: *J. Electrochem. Soc.*, 1963, **110**, p. 412

electron-hole pairs. The concentration of holes produced in this way allowed the dissolution reaction to take place in a uniform manner. The specimens were illuminated by perforating the cathode disc and shining a light through it.

A variety of electropolishing solutions has been suggested for both germanium and silicon. With the apparatus outlined above it was found that a solution containing 25 percent of glycerine in 0·1 percent KOH worked well. For silicon a dilute solution of HF, also with added glycerine was used as well. The results obtained were very good. The surfaces were first mechanically polished with 12 μm Al_2O_3 and then electropolished to remove the damaged layer. The smoothness of the surfaces was usually as good as that obtained by a conventional mechanical polishing technique finishing with 0·05 μm Al_2O_3.

Immediately prior to loading the semiconductor slice into the diffusion apparatus it is cleaned to remove any contaminants from the surface. The process usually involves ultrasonic cleaning in an organic solvent such as trichloroethylene and washing in distilled water. There are often several other steps and the cleaning process can become quite complex. An example of a full cleaning procedure for silicon has been given by Smith (1967).*

*Smith, A. M., 'Fundamentals of silicon integrated device technology' Vol. 1, Ed. Burger, R. M., and Donovan, R. P., (Prentice-Hall, 1967)

7.2.1 Open- and closed-tube methods

In the open-tube process, a long silicon tube is put into the furnace and a suitable carrier gas is passed through. The exit end of the tube is open to the atmosphere. A typical arrangement is shown in Fig. 7.2 in which the impurity source is in solid form and placed within the tube. A second furnace is used to heat the source to a temperature at which it has an appreciable vapour pressure. This will normally be less than diffusion temperature. The flow of the carrier gas then takes the vapour into the sample region. The samples are contained in a holder which can be pushed into the furnace area from the exit end of the tube when the diffusion experiment commences. A constant doping vapour is therefore kept in the vicinity of the samples: the experiment is concluded by withdrawing the sample holder. A ball of quartz wool is often placed upstream of the samples to introduce turbulence into the gas flow. Separate thermocouples are placed close to the sample holder and the impurity source to monitor the exact temperatures of source and samples.

There are a number of advantages to this system. The holder can accommodate many slices and is easily demountable. However, the thermodynamic parameters of such a system are usually not accurately known. There is also the point that the vapour of the semiconductor itself is continuously swept away by the gas flow, so the semiconductor is evaporating all the time, unless the surface is protected in some way. If the material has a high vapour pressure over itself at the diffusion temperature, this might lead to a serious loss of material.

In the closed-tube technique the prepared semiconductor sample and the diffusion source are initially placed inside a silica ampoule, open at one end with a constriction about half-way along (Fig. 7.3). The constriction is as small as possible, consistent with getting the material in, and the volume of the section containing the sample is usually in the range 2 to 15 cm^3. The open end of the crucible is attached to a vacuum apparatus and pumped down. When the pressure has dropped to about 10^{-6} torr, the tube is sealed off at the constriction. The sealing-off is best carried out using a gas torch with an oxygen-hydrogen flame. There is a certain amount of

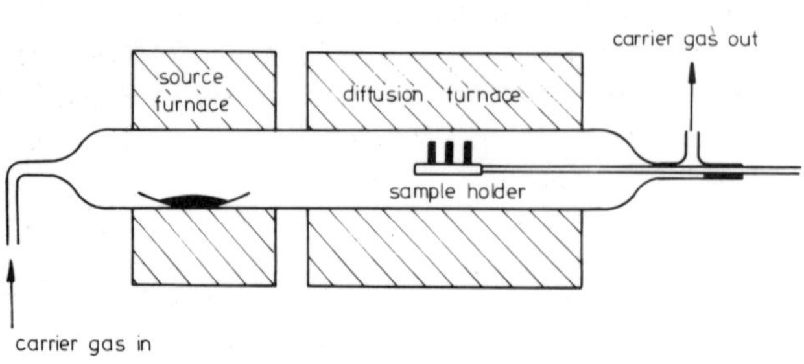

Fig. 7.2 Diffusion by an open-tube method

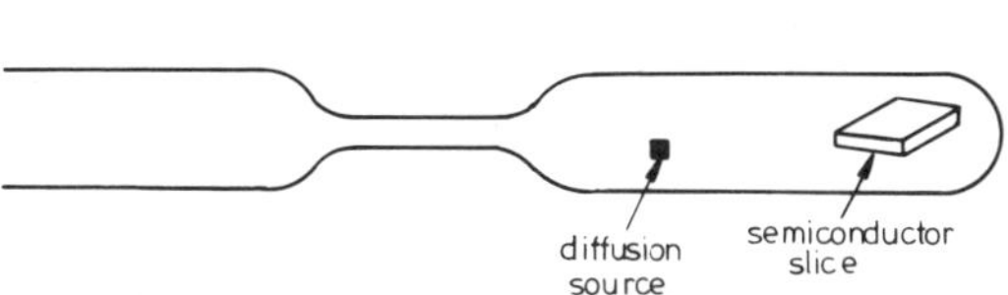

Fig. 7.3 Diffusion by a closed-tube method

skill involved, but it does come with practise: it is all too easy to knock a hole in the ampoule, letting in the air and ruining the experiment!

The sealed ampoule is then put into the diffusion furnace. Usually it is required that the whole of the ampoule should be at the same temperature: There must then be a constant temperature zone in the diffusion furnace that is longer than the ampoule. Sometimes it is required that the source and sample should be at different temperatures. In this arrangement, the slice is kept at the diffusion temperature and the source is kept at some other temperature, usually lower. In principle the arrangement has the advantage that the vapour pressure of the source can be altered by varying the second temperature, giving an extra freedom to the experimenter. In practise, it is done by having a temperature gradient along the length of the ampoule instead of a steady temperature. The ampoule is made fairly long, and the semiconductor is put at one end, with the source at the other. Such a system certainly works, but if the object of the experiment is to understand the diffusion process, it is probably not a good idea. The system is so far from equilibrium that normal theory, such as that described in chapter 3, cannot be used. The simplest way to carry out closed-tube diffusion is to have no concentration gradient along the ampoule. The situation can then be regarded as one in which everything inside the tube is a part of the system and equilibrium thermodynamics can then be applied. Even this is an approximation, of course, since a sample undergoing diffusion is not in equilibrium (see chapter 3).

At the end of the diffusion time the ampoule is removed from the furnace, quenched, and broken open. The sealing-off and breaking-open procedures make closed-tube diffusion troublesome and expensive. It is therefore not very suitable for large-scale production. However, it has already been pointed out that when diffusion mechanisms are studied, it is preferable to use a closed system. It also has the advantage of not sweeping away the vapour surrounding the semiconductor. Once the semiconductor has evaporated sufficiently to build up an appropriate atmosphere of itself, it does not evaporate any more. This is particularly important with some compound semiconductors. The system is also suitable for use in tracer experiments, in which a radioactive isotope is used for the source: it is hardly desirable to have radioactive materials going in a constant flow through an open-tube system.

Care must be taken that a large pressure does not build up inside the ampoule at diffusion temperature, causing an explosion inside the furnace. A relatively small piece of phosphorus, for instance, can look most inoffensive when the tube is loaded at room temperature prior to pumping down the ampoule. At 1000°C, however, it might be capable of building up a vapour pressure of several atmospheres in a small ampoule. It is therefore essential to calculate beforehand an approximate value for the pressure that will occur. This is especially important if radioactive sources are used: an explosion under these circumstances can be serious.

According to the theory of chapter 3, the ampoule will in general contain solid, liquid and vapour phases. In some cases no liquid phase forms and there is only a solid phase (the semiconductor slice) and a vapour phase consisting of semiconductor vapour plus source vapour. This amounts to saying that all of the added impurity has gone straight into the vapour phase. When this happens, another interesting effect occurs. During the course of the diffusion, more and more dopant atoms are incorporated in the semiconductor. Since these are taken directly out of the vapour phase, it follows that the vapour pressure must be falling all the time. As noted earlier, the surface concentration in the semiconductor is fixed by the external vapour pressure, so that must also fall throughout the experiment. The situation is therefore quite complicated and the diffusion profile should not be expected to have a simple form. The effect will only be important if the amount of dopant that enters the semiconductor is an appreciable fraction of the amount added to the tube at the start of the experiment. This is a point which should always be checked before advancing exotic theories to explain strange diffusion profiles.

7.2.2 Diffusion sources

Fig. 7.2 shows a solid source being used with the open-tube process. For diffusing phosphorus into silicon, for instance, anhydrous phosphorous pentoxide is a suitable source. It is interesting that although the P_2O_5 is the primary source, an intermediate reaction takes place between the P_2O_5 and the silicon, forming a layer, and it is from this layer that the diffusion proceeds. The point will be taken up again below. The arrangement can easily be modified to use a liquid source. This is done by bubbling the carrier gas through the liquid before passing it into the furnace. The carrier gas then takes a vapour of the dopant into the sample region, as before. The source furnace is no longer necessary, of course, but it may be necessary to heat the liquid to increase its saturated vapour pressure. Phosphorus can be diffused into silicon in this way using the liquid phosphorus oxychloride ($POCl_3$). A gas source can be used by simply mixing it with the carrier gas.

Solids, liquids and gases can be used as sources in closed-tube diffusion, although it is probably true to say that solids are most commonly used. Very often the source is simply a piece of the element to be diffused. The amount of source must be decided beforehand and weighed out. The required weight sometimes works out to be a rather small amount and handling difficulties arise. When this is so, it helps to remember that the thinner the material, the larger is the area. In extreme cases, it is a good idea to evaporate a thin film of the source material on to a piece of silica, which can easily be picked up with tweezers. It is fairly easy to prepare a source weighing just a few micrograms in this way, using silica dice a few millimeters square in area.

The diffusing material can also be applied directly to the semiconductor surface so that the diffusion is from a thin layer at the surface (see chapter 2). The impurity layer can be deposited by vacuum evaporation, electroplating or, sometimes simply by brushing a liquid on to the surface.

Many diffusion systems which, on the face of it, involve diffusion from the vapour phase do, in fact, work by diffusion from a surface layer. An example has already been given, namely the use of solid P_2O_5 as a source for phosphorus diffusion in silicon. The carrier gas takes the P_2O_5 vapour to the semiconductor but the diffusion does not occur directly from the vapour. A chemical reaction takes place between the silicon and the P_2O_5, forming phosphorus and silicon oxide, which combine in a glassy phosphate layer. It is from this layer that diffusion actually proceeds. Similarly when selenium or tellurium is diffused into GaAs in a closed-tube system, a glassy layer can form. One advantage of a glassy layer is that it forms a protection for the underlying semiconductor surface.

Layers of this type can be created rather more deliberately by depositing silicon oxide on the semiconductor surface prior to the diffusion process. The technique is not confined to diffusion in silicon: silicon oxide layers can be deposited on a number of semiconductors. The layer can be sputtered on, but it is more usual to employ a thermal decomposition process of an appropriate material, such as silane (SiH_4). A vapour of this material is passed over the semiconductor, which is heated. The silane decomposes, depositing silicon oxide on the surface. It is possible to introduce a vapour of the doping atoms during the decomposition process, so that the oxide layer contains the dopant. The silicon oxide then becomes the diffusion source, and the process is very similar to that described for P_2O_5, except that it is rather more controllable. Alternatively a pure oxide layer can be put down. The diffusion can then be carried out in one of two ways. In the first a preliminary diffusion is performed in which the oxide layer is doped. The specimen is then removed from the doping vapour and heated again, to allow the dopant to diffuse from the oxide into the semiconductor. A second technique, which is simpler, is to prolong the first diffusion so that the dopant is essentially diffusing through the oxide into the semiconductor. This is one of the cases solved mathematically in Chapter 2. The two methods give rise to slightly different diffusion profiles.

When oxide films are used in this way, the semiconductor surface is protected and this is an advantage from the point of view of device manufacture. It is also possible to achieve lower surface concentrations.

An ingenious technique, really suitable only for closed-tube work, is to use a sample of the semiconductor itself as the diffusion source, this sample being doped already with the desired impurity. The source material is crushed to a powder before being added to the diffusion ampoule. At diffusion temperature, some dopant atoms diffuse out of the source to provide an ambient vapour. Equilibrium between the source and vapour comes about very quickly because the former is in powder form and has a large surface area. Dopant atoms then diffuse into the semiconductor sample from the vapour. It is important to remember that not all of the dopant atoms diffuse out of the powder: an equilibrium is reached between the source and vapour in which the rate that atoms leave the powder equals the rate at which they return. Since the sample and powder are both in equilibrium with the same vapour, the concentration of doping atoms in the powder must at all times during the experiment be equal to the surface concentration in the sample.

After diffusion has proceeded for some time, the dopant atoms, which were initially all contained in the powder, are now distributed between the powder, the vapour, and the sample. If the amount of powder used is quite large, the fraction of dopant atoms leaving the source can be small. This implies that the doping of the powder at the end of the experiment is not very different to that at the beginning. Since this doping is also equal to the surface concentration of the sample at the end of the diffusion, the following procedure is suggested. A fairly large sample of the semiconductor, doped homogeneously to the desired level C_0 is chosen, and crushed. It is used as the source for diffusion into a sample of the same semiconductor. The surface concentration of the sample after diffusion is C_0, or just a little less. A good degree of accuracy in doping can be obtained by this technique.

If the diffusion source is a compound which undergoes a chemical reaction with the semiconductor, then it proves to be rather difficult to understand the fine details of the process. In a constant flow system with a compound source, so many things are happening at the same time; chemical reaction (which may take place in more than one stage), diffusion and, possibly, evaporation of the specimen. It is a much simpler situation if the semiconductor is in contact with a vapour of the element to be diffused, especially if the closed-tube technique is used. Once again, it comes down to what the point of the diffusion experiment is. If the object is to make devices, then the main criteria are those of cost and convenience, and it does not really matter if the experiment is complex, providing it works. If the reason for carrying out the diffusion is to study diffusion mechanisms, however, the most sensible course, as always, is to simplify the experiment as much as possible.

It should be clear from all this that there is a good deal of scope for the experimenter to choose the type of diffusion system and the form of source to be used. A comprehensive survey of the literature has been published (Sharma, 1970)* giving the relevant experimental details for a large number of experiments that have been carried out using different diffusants and semiconductors.

7.2.3 Box method

This method is really a variation of the closed-tube technique. Instead of sealing the tube under vacuum, it is merely closed with a mechanical seal so that the tube (or

*Sharma, B. L., 'Diffusion in semiconductors', *Trans. Tech. Publ.,* (Clausthall-Zellerfeld, 1970)

'box') is not quite gas tight. A typical arrangement, due to D'Asaro (1960)* is shown in Fig. 7.4. The tube is composed of two halves, held together by a platinum band. Since the arrangement lets in air, the diffusion takes place in an oxidising atmosphere, and the presence of a film of silicon oxide retains the high quality of the surface.

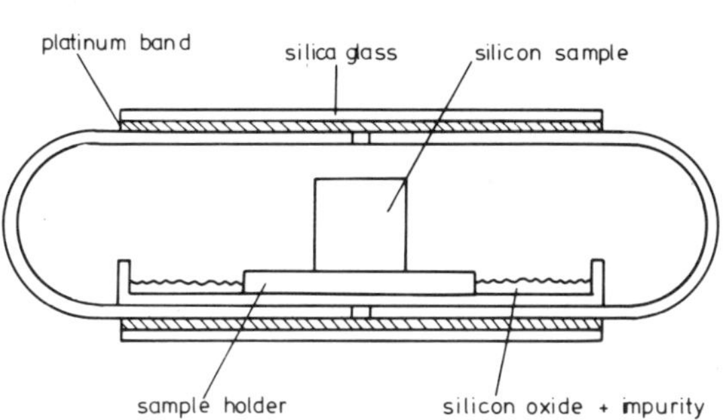

Fig. 7.4 Diffusion by the box method

Reproduced from D'Asaro, L. A.: *Solid State Elect.,* 1960, **1**, p. 3

The diffusion source is doped silicon oxide. At the diffusion temperature, an atmosphere of the dopant builds up in the box and equilibrates with the source silicon oxide and the layer of oxide on the silicon slice. Since they are both in equilibrium with the same vapour, the doping in the two lots of oxide becomes equal. This is very much the same technique as that described above in which a powder of the doped semiconductor is used as source. Diffusion into the silicon then takes place from the doped oxide layer.

The degree of sealing by the platinum band is fairly important. The enclosure must be sealed tightly enough to prevent a large loss of impurity vapour, but must allow oxygen to enter so that an oxide film forms on the silicon. For this special case of silicon diffusion, and possibly one or two other semiconductors, the box method has a lot to recommend it. It combines most of the advantages of the closed-tube method, with the added virtues of needing no sealing-off procedure and being easily demountable.

7.2.4 Oxide masking

It has been described how a doped oxide layer deposited on the semiconductor surface can be used as a diffusion source. Diffusion can also take place through an undoped oxide layer, giving a low surface concentration to the resulting profile in the semiconductor. However, it is fairly clear that if the oxide is thick enough and

* D'Asaro, L. A., *Solid-State Elect.*, 1960, **1**, p. 3

if the diffusion time is not too long. a negligible number of doping atoms will diffuse through the oxide and enter the semiconductor. The oxide is here acting as a barrier to diffusion.

The property is of importance in integrated circuit manufacture. When making circuits it is frequently necessary to diffuse dopants into certain highly localised regions of the semiconductor slice. This is achieved by covering the surface with a layer of oxide, and making holes at appropriate places. The remaining oxide film holds back the diffusion process, and dopant enters the semiconductor only where there are holes. This is very simple: the skilled part involves making the holes of exactly the right size and in exactly the right places. A photographic technique is used. A continuous film of oxide is first deposited on the semiconductor surface. It is then covered with a film of photoresist. Photoresists are materials which undergo marked changes when exposed to ultraviolet light: they 'harden' and their solubility in certain solvent systems is greatly decreased. A photographically produced mask is next placed over the photoresist film. The mask shows the required pattern of holes: in the areas where holes are wanted in the oxide layer the mask is opaque, elsewhere it is transparent. The whole arrangement is exposed to ultraviolet light and then etched in the appropriate solvent. The regions of photoresist not exposed to the light are still 'soft' and are dissolved away, leaving a pattern of holes that corresponds to the desired pattern. The sample is then etched again in a different solution that attacks the unprotected oxide, forming the pattern of holes. Yet another solution is now used to dissolve away the 'hard' photoresist, its job completed. Diffusion can then be carried out in the normal way.

The diffusions needed in the planar technology, as this type of manufacture is called, are often very precisely defined. It might be required to make a p–n junction only a micron or so below the surface of the semiconductor. A very accurate and reproducible technique must therefore be developed, especially if the devices are to be made on a large scale. The photographic masks have to be made to a high degree of precision. They are usually prepared initially at least ten times the final required size, as accurately as possible. They are then reduced to the correct size by photographic means. The preparation of an integrated circuit might involve many steps of the sort described above, each step using a mask with a different pattern.

7.2.5 2-stage diffusions

When making semiconductor devices, the diffusion is often carried out in two steps. The first is a 'normal' process using one of the techniques outlined above. In the second step, sometimes called the 'drive-in' stage, the sample is again taken up to diffusion temperature, but this time not in an atmosphere of the dopant vapour. The dopant diffuses out of course, lowering the surface concentration. Impurities also diffuse further into the semiconductor, so the net effect is to obtain a deeper diffusion penetration while at the same time decreasing both the surface concentration and the total number of atoms in the semiconductor. All of these features are desirable for certain device purposes. If there is an oxide layer on the semiconductor, as there is likely to be if the sample is silicon, the last effect will be less marked: most impurities diffuse much more slowly in silicon oxide than in silicon, so the oxide layer will seal the impurities into the semiconductor.

2-stage diffusions can also provide useful information in experiments carried out to study diffusion mechanisms. In general the diffusion coefficient and solubility of an impurity in a semiconductor will depend on the level of impurities already present in the material. The initial doping can be achieved in a preliminary diffusion, and it is instructive to follow the diffusion properties of the second impurity as a function of the concentration of the first. Probably the most useful 2-stage diffusion experiment, however, is one in which both diffusions involve the same element. This is the 'isoconcentration' technique, briefly mentioned in Chapter 6.

If the diffusion coefficient is a function of concentration, Fick's second law becomes

$$\frac{\partial C}{\partial t} = \frac{\partial}{\partial x}\left\{D(C)\frac{\partial C}{\partial x}\right\} \tag{7.1}$$

Here the diffusion profile is not a simple function. The problem, from the experimental point of view, is usually to find the function $D(C)$ from the diffusion profile. A simpler approach is the following. An initial diffusion is carried out using a closed-tube method, in which the semiconductor is doped homogeneously with the impurity. Let the concentration be C'. The specimen is removed from the tube and cleaned. A second experiment is then carried out, this time using a radioactive isotope and a shorter diffusion time. The experiment is in all other respects identical to the first, however. The same size of diffusion ampoule is used, and the weights of materials within it are identical. At the end of the diffusion a profile of the radioactive impurity in the semiconductor is plotted.

Now because the same amounts of material are used in both experiments, the second one takes place at equilibrium. The semiconductor is surrounded by the correct vapour pressure of dopant and so there is no net exchange of atoms between solid and vapour: the same number of atoms join the vapour from the solid as leave it. Although the semiconductor cannot tell the difference between the atoms joining it from the vapour and those leaving it, we can, since the former are radioactive. The ratioactive atoms diffuse in a crystal that is always homogeneously doped to a level C', and eqn. 7.1 becomes

$$\frac{\partial C}{\partial t} = D(C')\left(\frac{\partial^2 C}{\partial x^2}\right) \tag{7.2}$$

This equation has the standard error function complement solution for the experiment described. It gives a value of D for the specified concentration C'. The experiment can be repeated a number of times for different values of C', and the function $D(C)$ is determined.

The isoconcentration technique is an example of diffusion taking place when there is a concentration gradient of the radioactive isotope but not a net gradient of the element. According to Section 5.1, the measured diffusion coefficient is the tracer coefficient, D^*. Strictly, this is different to the coefficient measured in a nonisoconcentration experiment, D, which is given by eqn. 5.6 as

$$D = D^* \left(1 + \frac{d \ln \gamma}{d \ln N}\right)$$

If the solution of the diffusing atom in the crystal is dilute, however, then Henry's law holds, and the second term in the bracket is zero. The two coefficients are then equal.

7.3 Evaluation of diffused layers

Having carried out a diffusion, the experimenter usually wishes to know what exactly he has done. The important parameters which are required from a diffusion experiment are the surface concentration and the diffusion coefficient. There are several techniques available for obtaining this information, the most suitable one depends on the details of the experiment.

7.3.1 *p–n* junction method

This technique is suitable for an experiment in which a p-type dopant is diffused into a homogeneously doped n-type substrate (or vice versa). The important thing is that a p–n junction should be formed. It is necessary to know the original n-type doping of the specimen. This can be found by carrying out a routine electrical measurement on the specimen before diffusing it. It is also necessary to measure the depth of the p–n junction after the diffusion, and to assume that the form of the profile is one of the 'simple' forms given when the diffusion coefficient is a constant.

Take, for instance, the diffusion into the sample from an 'infinite' vapour source, i.e. the amount of material in the vapour phase is large compared to the amount that enters the sample. The surface concentration is constant throughout the experiment and, according to chapter 2, the profile has the form

$$C = C_0 \operatorname{erfc}\left(\frac{x}{2(Dt)^{1/2}}\right) \tag{7.3}$$

where C_0 is the surface concentration and D is constant. Suppose the sample was originally n-type, doping C_n, and the p–n junction appears at a distance b below the surface. Then at this special point, eqn. 7.3 becomes

$$C_n = C_0 \operatorname{erfc}\left(\frac{b}{2(Dt)^{1/2}}\right) \tag{7.4}$$

Since at the junction $n = p$ and the assumption is made that all donors and acceptors are ionised.

A single pair of values for C_n and b will not give C_0 and D, but if the same experiment is repeated for a second sample, originally doped to a level C_n', then these parameters can be found. Suppose the junction depth in the second experiment is b'. Then taking the ratio of two expressions of the type eqn. 7.4, we have

$$\frac{C_n}{C_n'} = \frac{\operatorname{erfc}\left(\dfrac{b}{2(Dt)^{1/2}}\right)}{\operatorname{erfc}\left(\dfrac{b'}{2(Dt)^{1/2}}\right)} \tag{7.5}$$

In which D is the only unknown. Having found D, C_0 can be calculated. Experimentally this is much harder than it sounds. The two junction depths are not likely to be very different, so it is necessary to measure them accurately to get good results. It also helps to have the two original doping levels very different, $C_n >> C_n'$, say.

If $C_n << C_0$, an approximation (Dunlap, 1954)* can be made. Eqn. 7.4 shows that under these circumstances the erfc term is very small, and we can make use of the fact that for this condition,

$$\operatorname{erfc} y \simeq \frac{1}{\pi^{1/2} y} \exp(-y^2)$$

Substituting this into eqn. 7.4 and simplifying gives the approximate result

$$b \simeq 2(Dt)^{1/2} \left(\ln \frac{C_0}{C_n}\right)^{1/2} \tag{7.6}$$

Since the ratio C_0/C_n enters only by its logarithm, the expression is not very sensitive to its value. Suppose the ratio is 10^4, then eqn. 7.6 comes to

$$b \simeq 6(Dt)^{1/2} \tag{7.7}$$

underlining the point made in chapter 2 about the length $(Dt)^{1/2}$ being the 'natural' unit of measurement in diffusion work. Thus an approximate value for D can be found from a single diffusion.

As an alternative to the 'infinite source' set of boundary conditions, one can consider the 'constant source' case, in which a thin layer of diffusing material is plated on the surface of the specimen. The diffusion takes place from this thin layer, the amount of diffusing substance in the specimen staying constant. Under these conditions, the surface concentration falls throughout the experiment and the profile at time t was shown in chapter 2 to be

$$C = \frac{\alpha}{(\pi Dt)^{1/2}} \exp\left(\frac{-x^2}{4Dt}\right) \tag{7.8}$$

*Dunlap, W. C., Phys. Rev., 1954, **94**, p. 1531

where α is the number of impurity atoms deposited on unit area of the sample. If the surface concentration C_0 is much larger than that at the junction C_n, say $(C_0/C_n) > 10^4$, the position of the p–n junction becomes relatively insensitive to the precise value of C_n. This is because eqn. 7.8 is falling so rapidly with increasing x at that point. For this condition, a similar argument to that given above for the erfc solution yields

$$b \simeq 5(Dt)^{1/2} \tag{7.9}$$

It is important to note that to achieve the constant source condition in an experiment it is necessary to prevent any of the diffusing layer evaporating off the surface. If evaporation occurs, then the amount of diffusing material is not constant throughout the experiment. Thus simply plating a layer on to a slice of semiconductor and bringing the specimen up to diffusion temperature does not usually give constant source conditions: most of the layer goes straight into the vapour phase. If this happens in a closed-tube experiment, it gives rise to conditions more like 'infinite source' than 'constant source' and eqn. 7.3 is probably more appropriate than eqn. 7.8. However, in view of the similarity of the solutions for the two cases (compare eqn. 7.7 with eqn. 7.9), the point is not important unless high accuracy is required. Then, one would probably not use the p–n junction method anyway.

The convenience of the method depends on the ease with which the p–n junction depth can be located. The simplest method is to use a chemical etch which will reveal the junction. It has already been noted that electrons and holes can take part in the dissolution reaction when a semiconductor is being etched, and this means that some etches behave differently with n- and p-type semiconductors. An etch consisting of concentrated HF containing about 0·5 percent HNO_3, for instance, darkens p-type silicon, but not n-type silicon (Fuller and Ditzenberger, 1956).* The procedure is therefore to section the diffused specimen perpendicular to the top surface and to etch the cross-section in an appropriate liquid. The junction is revealed when the specimen is viewed under a microscope, and its depth can be measured. If the junction is just a few microns below the top surface, then the length to be measured is likely to be inconveniently small. The length can effectively be magnified if the specimen is bevelled at some small angle to the surface, rather than being section at 90°. The idea is shown in Fig. 7.5. The distance to be measured is now l rather than d, a magnifying factor of cosec θ, i.e. if $\theta = 5°$, a factor of about eleven.

Some etches work by preferentially attacking the junction area rather than by etching the p and n parts differently. This gives rise to a dark line on the etched surface corresponding to the junction. A difficulty which sometimes arises is that while the etches are sensitive to p–n junctions, they also tend to be sensitive to other features as well. If a diffusion profile demonstrates a sharp drop in

* Fuller, C. S., and Ditzenberger, J. A., *J. Appl. Phys.*, 1956, **27**, p. 544

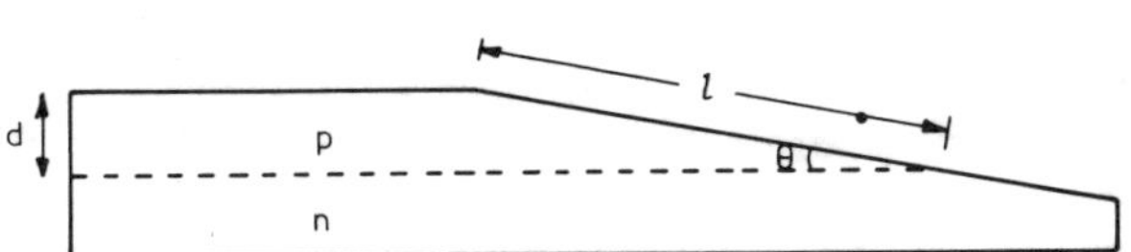

Fig. 7.5 Increasing effective junction depth by bevelling a semiconductor slice

concentration at some depth, for instance, this is also likely to be shown up as a line. An embarrassingly large number of lines can therefore be obtained – perhaps three or four – and it is necessary to decide which of them corresponds to the junction.

A second method which has been used to determine the position of the junction utilises the fact that the Seebeck coefficients for *n*- and *p*-type materials are of opposite sign. The thermoelectric e.m.f.'s obtained from the two sides of the junction are therefore in opposite directions. Fig. 7.6 shows an arrangement which can detect a junction using this effect. The specimen is bevelled, as described above. A thermocouple is made with the cold junction a permanent connection to the *n*-substrate, and the hot junction a heated probe which can scan across the junction. This probe should preferably be mounted on a micromanipulator. The thermoelectric e.m.f. is detected by the galvanometer. As the probe crosses the junction a sharp change is observed in the meter reading.

The attraction of the *p–n* junction method of evaluation is that it is easy to do on a routine basis, and gives an approximate value of diffusion coefficient for a very small amount of experimentation. However, it must be used with caution because of the large number of implicit assumptions. The experiment actually tells us that

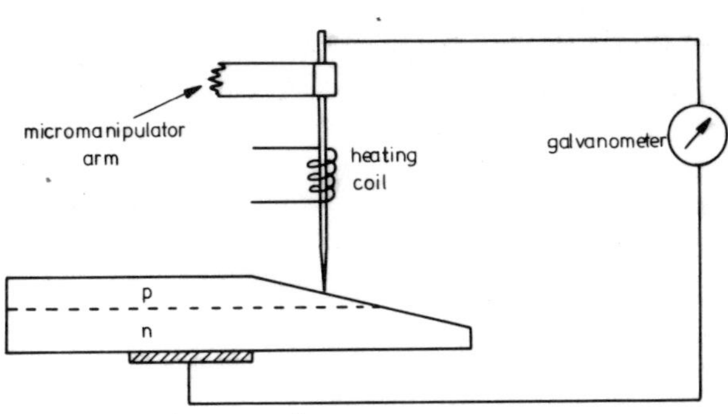

Fig. 7.6 Location of a *p–n* junction using the thermoelectric effect

at a certain distance b the concentration of holes is equal to the concentration of electrons C_n, i.e. we are given a single point on the diffusion profile. The situation could be improved, of course, by doing the experiment many times for different C_n. The whole diffusion profile could then be plotted. However this is a good deal more trouble and expensive on material, and if one is prepared to do this much work it is worth considering a radio-tracer technique. To extract information from a single result, it is necessary to make a prediction about the form of the profile. It is invariably assumed in doing this, that the diffusion coefficient is a constant, and the diffusion process is simple. We have seen in previous chapters that this can be a rash assumption when dealing with semiconductors. Finally, it is assumed that all donors and acceptors are ionised, so that at the junction the number of donors is equal to that of acceptors. It has been shown that this is not always the case: the point will be taken up later in the chapter.

7.3.2 4-point probe method

This is a very useful technique which can be used to plot diffusion profiles. It is also employed for determining resistivity of thin homogeneously doped samples such as epitaxial semiconductor layers. The method is valid for diffusions of n-type dopant into p-type substrates, or vice versa, i.e. a p–n junction must be formed. Similarly, when used to characterise epitaxially-grown layers, a layer must be grown on a substrate of opposite type. The method is essentially the same as one used in geophysics for measuring the resistivity of the earth.

The arrangement is shown in Fig. 7.7. Four equally spaced probes are applied to the semiconductor surface. The distance between them s is usually of the order of 1 mm or less. A current I of the order of the milliamps, is passed through probes 1 and 4 and the voltage V is measured between probes 2 and 3. The voltage might typically be in the millivolt range. The relationship between V, I, and the conductivity depends on the geometry of the specimen. It has been calculated for most arrangements which are likely to arise in semiconductor work (Smits, 1958).* We shall first of all consider the simplest case, the semi-infinite specimen, and then go on to indicate how the method is used for determining diffusion profiles. It is assumed that any minority carrier injection from the probes into the semiconductor is negligible, and that the surface on which the probes rest is flat with no surface leakage.

A 'semi-infinite' specimen stretches to infinity in the $\pm x$, $\pm y$ and z directions (see Fig. 7.7a). This may be taken to mean a crystal which is large compared to the separation between the probes. The problem of finding a value for the conductivity in terms of the measured voltage and current is solved by making use of the fact that the differential equations for ohmic flow are identical with those of electrostatics (and with those of heat flow, for that matter). Associated with the flow of current through probes 1 and 4 is a potential distribution within the sample. This same distribution could be achieved by having a positive charge at the

* Smits, F. M., *Bell System Tech. J.*, 1958, **37**, p. 711

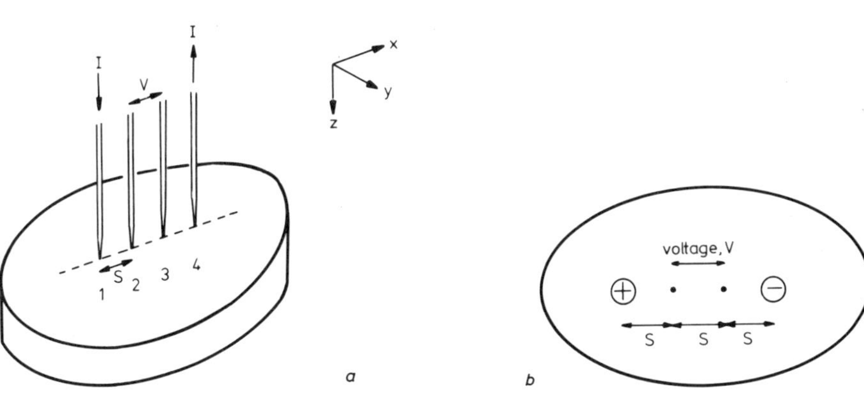

Fig. 7.7 *a* 4-point probe technique

b electrostatics analogue for the 4-point probe method

point where probe 1 touches the surface and an equal negative charge at the point on the surface where probe 4 touches. Call these charges $+q$ and $-q$ (see Fig. 7.7b). As far as the potential distribution is concerned, the first problem, involving current flow, and the second in electrostatics, are identical. The link between the two models is forged by deriving a relationship between I in the first and q in the second.

The potentials at probe 2 due to the two charges are $q/4\pi\epsilon s$ and $-q/8\pi\epsilon s$ respectively, where ϵ is the permittivity of the material. The potential difference between points 2 and 3 therefore comes to

$$V = \frac{q}{4\pi\epsilon}\left(\frac{1}{s} + \frac{1}{s} - \frac{1}{2s} - \frac{1}{2s}\right) = \frac{q}{4\pi\epsilon s} \tag{7.10}$$

The electric field associated with the potential distribution is given by

$$\mathscr{E} = \frac{q}{4\pi\epsilon r^2} \tag{7.11}$$

Now return to Fig. 7.7a and consider a hemispherical surface of infinitesimal radius r centred on one of the current probes. Applying Ohm's law, $J = \sigma\,\mathscr{E}$, to the current flowing through this surface

$$\frac{I}{2\pi r^2} = \sigma\,\frac{q}{4\pi\epsilon r^2} \tag{7.12}$$

where σ is conductivity.

The relationship between Figs. 7.7a and 7.7b is therefore given by

$$q = \frac{2\epsilon I}{\sigma} \tag{7.13}$$

which, when substituted in eqn. 7.10 gives

$$\sigma = \frac{1}{2\pi s} \cdot \frac{I}{V} \tag{7.14}$$

for the semi-infinite specimen.

The next most interesting problem is the one for which the specimen is still effectively infinite in the $\pm x$ and $\pm y$ directions, but is thin in the z direction. This applies to the measurement of thin semiconductor layers. Let the thickness of the layer be w. The advantage of equating the problem to its electrostatics equivalent now becomes apparent. The potential distribution within the layer can be calculated using a standard electrostatics technique called the 'method of images'. The bottom surface of the slice interferes with the potential distribution, and it can be shown that this interference can be allowed for by treating the top and bottom surfaces as mirrors in which the original $+q$ and $-q$ charges are multiply reflected. This is shown in Fig. 7.8, in which it is assumed that the surfaces are nonconducting: a conducting surface reflects an image of opposite sign. The procedure for finding V is then similar to that described above. The only difference is that the potentials at points 2 and 3 are due not just to the original charges at points 1 and 4, but to all the virtual images as well. When the summation is carried out (Uhlir, 1955),* eqn. 7.14 becomes

$$\sigma = \frac{I}{4{\cdot}5w} \cdot \frac{I}{V} \tag{7.15}$$

where it has been assumed that $s > 2w$. If this condition is not satisfied, the figure of 4.5 is not quite correct: the correction factors are given in the article by Uhlir (1955). Making the sample finite in the x and y directions introduces more surfaces, and the virtual images due to these surfaces must be taken into account. The problem becomes more complicated, but is the same in principle.

We can now return to a diffused specimen containing a *p–n* junction. Suppose an *n*-type dopant has been diffused into a *p*-type specimen from the top surface. The concentration of donors decreases with distance from the top surface, and the plane of the *p–n* junction is parallel with the surface. Because a *p–n* junction region is depleted of charge carriers, it acts as an insulating layer in this arrangement. When the 4-point probe measurement is made, therefore, none of the current flows into the *p* region below the junction. It follows that it is necessary only to consider the diffused *n*-type region in any calculations. The best way to think of the layer is as a large number of sublayers piled on top of each other, each

* Uhlir, A., *Bell System Tech. J.*, 1955, **34**, p. 105

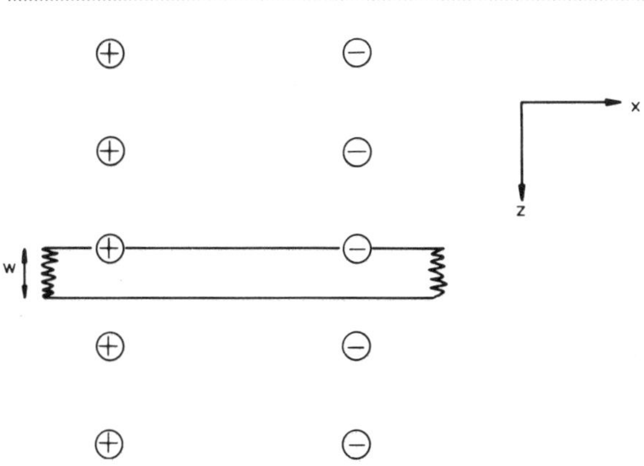

Fig. 7.8 Application of the method of images to a semiconductor slice

with a different resistance because of the changing electron concentration. The resistances of these sublayers are in parallel with each other, from the point of view of the current flow between the probes.

The profile is plotted by carrying out a 4-point probe measurement on the surface, then lapping off a thin layer and doing it again. The amount removed might typically be a few microns. The process is repeated until the p–n junction is reached. We assume the specimen is an infinite slab of thickness w, as in Fig. 7.8, so that eqn. 7.15 is appropriate. It is convenient to work in terms of surface conductance G rather than conductivity σ. The two are related by

$$G = w\sigma = \frac{1}{4{\cdot}5} \cdot \frac{I}{V} \tag{7.16}$$

where w is now the effective thickness of the specimen, i.e. the distance between the top surface and the p–n junction.

Consider two successive measurements on the slab, the first taken when the thickness is w and the second taken when a thin layer δw has been lapped off. Let the probe currents and voltages be V_1, I_1 and V_2, I_2, respectively. If the resistance of the slab in the first place is R_1 and after lapping, it is R_2, and if the resistance of the layer that was removed is R, then

$$\frac{1}{R} = \frac{1}{R_1} - \frac{1}{R_2}$$

because the resistances are in parallel. It follows from this, and from the definition of surface conductivity, that

$$G = G_1 - G_2$$

i.e.

$$\sigma(w)\delta w = \frac{1}{4{\cdot}5}\left(\frac{I_1}{V_1} - \frac{I_2}{V_2}\right) \tag{7.17}$$

The conductivity of the removed layer $\sigma(w)$ can therefore be calculated.

The experiment is repeated, removing further layers, and the variation in conductivity with depth $\sigma(z)$ can be plotted. This is related to the variation in electron concentration by the usual formula for conductivity:

$$\sigma(z) = n(z)B(z)e \tag{7.18}$$

where $B(z)$, the mobility, is also a function of distance below the surface, because mobility depends on electron concentration. For many semiconductors, data is available on the variation of mobility with electron concentration, i.e. the function $B(n)$ is known. It cannot be substituted into eqn. 7.18, since the equation contains $B(z)$, not $B(n)$. An iterative process (Fuller and Ditzenberger, 1956)* is used in which a guess is made at the value of n. The corresponding value of B is found from $B(n)$ and is substituted into eqn. 7.18. Using the experimental value of σ, a second value for n can be found. If this does not correspond to the original guess, a better estimate is made, and the cycle is repeated. A simpler method is to treat the mobility as a constant, taking some average value. The electron concentration can then be calculated directly from eqn. 7.18. This is a good deal less accurate, of course. Having found n, it is usual to assume that all donors and acceptors are ionised, and put

$$n = C_D - C_A$$

where C_A is the initial acceptor doping of the semiconductor.

If the specimen is not masked during diffusion, dopant diffuses in from the edge and bottom surfaces. For the experiment to correspond to the above theory, therefore, it is necessary to remove the diffused layer from the edges and lower surface. Care must also be taken when using the technique to have ohmic contacts between the current-carrying probes and the semiconductor surface. This can often be achieved by abrading the surface. A mechanically polished surface is usually what one has, anyway, since mechanical polishing is the most convenient way of removing thin layers. If the contact to an abraded surface is nonohmic, an ohmic contact can often be formed by passing a fairly high current through the outer contacts for a short time.

When polishing off the layers it is important to lap at 90° to the surface, and not at some small angle to this, otherwise inaccurate results are obtained. It is also necessary to measure the amounts abraded from the surface, i.e. the distances δw. One of the best ways is to use an optical interference technique in which the polished surface is compared to some reference surface. Differences in level can be measured to within about 0·2 μm in this way. Another simpler method is to

* Fuller, C. S., and Ditzenberger, J. A., *J. Appl. Phys.*, 1956, **27**, p. 544

measure the surface area of the sample, and weigh it before and after lapping. If the sample is of a reasonable size, the difference in weights can easily be measured on a sensitive balance, e.g. for a semiconductor of density 5 gm cm^{-3} and a specimen of 1 cm^2 in area, a 1 μm slice weighs 0·5 mg.

Another method of layer removal, specific to silicon, is to use anodic oxidation of the surface. A layer of oxide is formed on the surface by making the silicon slice the anode of an electrolytic cell (Schmidt and Michel, 1957).* The colour of the layer depends on its thickness, which can therefore be determined using calibration charts. The oxide is then removed by a chemical etch and a new surface is revealed for the next 4-point probe measurement. Very thin layers of just a few hundred Angstroms can be removed in this way. The method is especially useful for evaluating shallow profiles.

Under certain circumstances it is possible to compute a relationship between the average conductivity of a diffused layer, and the surface concentration. It is necessary for a p–n junction to be formed by the diffusion. The average conductivity, $\bar{\sigma}$, is found using the 4-point probe and substituting in eqn. 7.15. The thickness w in eqn. 7.15 is, here, the thickness of the diffused layer. It is also required to know the mathematical form of the diffusion profile and to have experimental data showing how electron and hole mobilities vary with doping. This information is available for some semiconductors and curves of surface concentration plotted against average conductivity have been computed. The method is described in more detail in Section 9.2, in which examples of curves calculated for silicon and germanium are also given.

The 4-point probe method is a simple and useful technique, and it is therefore widely used. It must be remembered that the information it gives directly concerns the electron and hole concentrations, not the atom concentrations.

7.3.3 Radiotracer method

The most direct way of finding the distribution of an element in a solid is to label it by using a radioactive isotope of the element. It is also probably the most expensive way, both in terms of time and equipment. Consideration must also be given to the health hazard of working with radioactive materials. Regular checks must be made on monitored workers, and disposal of radioactive waste must be carried out with great care.

In essence the experiment is simple. A closed tube diffusion is carried out using a radioactive source. The semiconductor slice is taken from the tube at the end of diffusion, and a thin layer is removed from the surface. The radioactivity of the layer is counted and compared with that from a standard piece of the doping element. The standard must be of the same radioactive level as that used in the diffusion, and be of known weight. The weight of the element in the layer can then be calculated, and if the dimensions of the layer are known, the mean concentration of the element in the layer can be found. Further layers are removed,

* Schmidt, P. F., and Michel, W., *J. Electrochem. Soc.*, 1957, **104**, p. 230

and the procedure is repeated. In this way the diffusion profile of the element in the semiconductor is plotted. The technique is very much the same as that used when plotting profiles using the 4-point probe. There is an implicit assumption here, of course, that the crystal cannot tell the difference between the isotope and the nonradioactive element, so that they both have the same diffusion coefficient.

In plotting the profile it is assumed that the measured mean concentration of the removed layer corresponds to the concentration at the centre of the layer. This is strictly true only for a linear profile, but if the layer is thin, the approximation is not too bad. In practise, the criterion for the thickness of the removed layer is that it should not be so thick that there is a large difference in concentration between its top and bottom surfaces.

The radioactive emissions of interest are alpha, beta and gamma rays. Alpha decay is the spontaneous emission of an alpha particle, which is the nucleus of the ^{4}He atom. It is mostly found in the heavier elements, of atomic number greater than about eighty. For this reason, it is not often encountered in semiconductor diffusion work. The beta particle is a high-speed electron emitted from a nucleus of a radioactive atom. They are given out in a continuous energy spectrum, covering the range from zero to some maximum value, with an average of about one third of the maximum. It turns out that this maximum energy is characteristic of the particular isotope concerned. Gamma rays are electromagnetic radiation, which therefore react with matter in a quantised manner as photons. The wavelength of the radiation is very small, i.e. less than for x rays. The radiation due to a given decay is monochromatic, and no two gamma-emitting elements give out photons of the same energy.

The isotope might emit one or more of these rays, at varying energies. Since the radioactive source is continuously giving out energy, it becomes steadily weaker with time, i.e. the number of emissions per second decreases. This is quantified by quoting a halflife for an emission, which is defined as the time required for the strength of a source to decline to half its power. The halflives of different emissions vary between a fraction of a second, and many years. In a radiotracer experiment it is desirable for the halflife to be longer than the duration of the experiment: if it is shorter, then each individual count must be checked against the standard. There is also the possibility that the isotope might decay below the limit of detection before the experiment is completed.

The types of radioactive counters commonly encountered in this type of work are:

(*a*) Gas-filled detectors: the Geiger-Muller tube is of this kind. The detector consists of a glass tube containing a gas. There is a pair of electrodes in the tube, between which a voltage is applied. When a radioactive particle enters the tube, it ionises the gas and the resulting ions and electrons migrate to the appropriate electrodes under the action of the applied field. A pulse of current therefore appears every time a radioactive particle enters the gas and the number of pulses per second is a measure of the number of particles entering the tube every second. The device gives a good signal, but it is difficult to use it to discriminate between two different types of radioactive emission.

(*b*) Scintillation counters: at the centre of this counter is a crystal which fluoresces when it absorbs radioactive emissions. The most widely used crystal is sodium iodide, which absorbs gamma rays efficiently, and can be prepared in the form of large single crystals (greater than about 20 cm in diameter). The fluorescence is detected by a photomultiplier tube placed in contact with the crystal. Consider what happens when a gamma-ray photon enters the crystal. Electron-hole pairs are created within the NaI, which recombine and create photons. The energy of these photons is characteristic of the NaI band structure, and is very much less than the energy of the original gamma photon. A gamma photon therefore creates a large number of secondary photons: one might say that a single large energy photon is turned into a number of small energy photons by the crystal. Now this all happens very quickly and, in effect, all of these secondary photons hit the photomultiplier face together, and a pulse is measured at the output of the photomultiplier. This happens every time a gamma photon is absorbed by the crystal so, once again, the number of pulses per second recorded by the counter is a measure of the number of radioactive particles entering it. A further useful facility is available using this system, however. A given isotope emits monochromatic gamma rays, i.e. photons of a single energy. The size of a pulse is dependent on the number of secondary photons which, in turn, is dependant on the photon energy of the original gamma photon. Thus if we have two radioactive isotopes emitting gammas of different photon energies, the counter will give out pulses of two different heights. It is a fairly simple matter to arrange an electronic counting system which will 'see' pulses of one height, but not of another. It is therefore possible to tune the system to count the pulses from the two isotopes separately. In principle, this facility gives rise to a whole range of possibilities for diffusion work. One could study the interaction of a pair of diffusing elements, for instance, if they were both gamma emitters, by carrying out an experiment in which they diffuse together. The diffusion profiles could then be plotted separately. This type of experiment does not seem to have been tried yet.

The technique of separating-out different emissions is more difficult if the emissions are beta rays. As already mentioned, the energy spectra of the rays, although characteristic of the isotopes concerned, are not monochromatic and there is therefore a tendency for two to merge into one another. The reader interested in the details of radioactive counting systems is referred to one of the several books on the subject (Kruger, 1971).*

Let us return to the diffusion experiment. The accuracy of the plotted profile

* Kruger, P. 'Principles of Activation Analysis' (Wiley-Interscience, 1971)

will depend very largely on the care with which the lapping and counting procedure is carried out. There are two straight-forward ways to remove thin layers from a semiconductor surface. The surface can be abraded. Here care must be taken to mechanically polish at right angles to the surface: a bevel of just a few degrees will give inaccurate results, since the penetration depth of the profile is likely to be only a hundred microns or so at most. The removed material must all be collected together and put into the counting system: this is not easy, since mechanical polishing tends to be a messy process. Alternatively, layers can be chemically etched from the surface. This method has a lot to recommend it. A simple technique is to carry out the etching in a small polythene container, a few cm^3 in volume, and to wash the sample in water contained in a second, similar, container. The two containers are then put into the counting apparatus, and one can be reasonably sure that no radioactivity has been lost. It is interesting that the second container can show as much as 10 percent of the total radioactivity, due to the volume of etch hanging on to the specimen by surface tension when it is removed. It is a good thing to have every sample in exactly the same physical form, from the point of view of reproducibility. The standard sample, also can be dissolved in etch in one of the containers, and this gives a good 'rigid' experimental technique.

There are disadvantages to the use of etches for removing layers. It has been mentioned already that it is difficult using etching to obtain a very flat surface: etch pits and other crystallographic features usually develop. In addition, all etches eat away preferentially at corners, so the edges of the sample become rounded. This means that for the first few layers at least, more material is removed from a thin anulus round the perimeter of the surface than from the rest. Whether or not this is important depends on the surface area: edge effects play a greater part in a specimen of small surface area than one with a large surface.

For the special case of silicon, the method of anodic oxidation can also be used for layer removal, as described for the 4-point probe technique.

So far it has been assumed that the diffusion takes place from the flat surface of the semiconductor slice. Unless the semiconductor is masked during diffusion, however, dopant will also diffuse in through the edges of the sample. Whether or not the amount diffusing in this way is appreciable, with respect to that going in through the large faces, depends on the size of the specimen. Take for instance, a slice of dimensions 1 cm x 1 cm x 0·05 cm. The area of the flat faces is 2 cm^2, and that of the edges 0·2 cm^2, suggesting that it is necessary to make some allowance for the edge effect, but we need not worry too much about errors in the allowance, since they constitute a second-order effect. If layers are being removed by abrasion, the best way is to remove an amount of material from the sides of the slice deeper than the estimated depth of diffusion. We are then only looking at dopant which has diffused in from the flat surfaces. When using etching it is probably best to add the area of the edges to that of the flat surfaces, and assume that the etch removes an equal layer from all over the specimen. Allowance can be made for the fact that as etching proceeds, and further layers are removed, the slice gets thinner, so the surface area of the edges gets smaller.

Having plotted the diffusion profile, the experimenter will want to know if it follows some simple (or complicated!) mathematical function of the type derived in Chapter 2. If this can be done, then some light will be thrown on the value of the diffusion coefficient of the dopant in the semiconductor and also, hopefully, on the diffusion mechanism. What is required, then, is to fit a set of experimental results giving concentrations C at different depths x to a mathematical function, $C(x)$. To do this in a credible way, it is essential to cover a large range of concentration in the experimental results. The lowest concentration measured should ideally be at least a factor of one thousand below the highest concentration measured. If relatively few experimental points are taken and they cover, say, only half an order of magnitude in concentration values, then by the time experimental error is allowed for, it is usually possible to fit an error function curve, or a straight line, or indeed any function that fits the experimenter's current theory. Fellow workers tend to be sceptical about such results.

Since a large range of concentrations must be measured, it follows that a wide range of radioactivities is counted. It is bad experimental practise to compare an unknown sample with a standard which differs from it by a large amount, and it is wise to have several standard samples, covering the whole range of radioactivity which is encountered in the experiment.

7.3.4 Comparison of electrical and radiotracer techniques

It has already been pointed out that the main difference between the electrical and radiotracer techniques is that the former gives charge carrier concentrations, whereas the latter gives atom concentrations. If all the donors and acceptors in the semiconductor are ionised, this amounts to the same thing. Very often they are all ionised and there is very little difference between results obtained using the two types of technique. An investigation of the diffusion of antimony into germanium, for instance, compared radiotracer measurements with the p–n junction technique (Dunlap and Brown, 1952).* The results agreed to within 10 percent. When p type InP is investigated, however, a quite different conclusion is drawn. Galvanov, Metreveli, Siukaev and Starosel'tseva (1969)† carried out conductivity and Hall effect measurements on InP homogeneously doped with zinc or cadmium. The material was prepared by prolonged diffusion and was made as heavily doped as possible. They came to the conclusion that the density of holes never exceeded 10^{19} cm^{-3}. At the same time the concentration of atoms resulting from the diffusion of radioactive zinc was 10^{20} – 10^{21} cm^{-3}, i.e. in heavily doped material the true concentration of zinc atoms exceeded the hole density by one or two orders of magnitude.

There are a number of reasons why the number of doping atoms might not correspond to the number of charge carriers. Some atoms might exist in a neutral state not giving rise to an electron or hole, or in an inactive state as part of a

* Dunlap, W. C., and Brown, D. E., *Phys. Rev.*, 1952, **86**, p. 417

† Galvanov, V. V., Metreveli, S. G., Siukaev, N. V., and Starosel'tseva, S. P., *Sov. Phys. – Semiconductors*, 1969, **3**, p. 94

precipitate, or similar feature. A fraction of the atoms might also exist interstitially, in which case they can become dopants of a different kind. An example of this last phenomenon is zinc in the III–V semiconductors. It can occur as an acceptor when on the group III site, but also as a donor when on an interstitial site. There are many other possibilities.

One of the most convincing demonstrations of a difference between carrier concentration and dopant concentration is given by the diffusion of sulphur in GaAs (Young and Pearson, 1970).* Sulphur, being group VI, is a donor in GaAs, so the electron concentration profile was measured and compared to the radiotracer profile. In these experiments the electron densities were measured by an ingenious optical technique, which relies on the fact that the optical reflection spectrum of a semiconductor depends on the electron carrier concentration. The position of the plasma reflection minimum was observed and, by reference to an appropriate calibration curve, the electron density was found. The reflection experiment was carried out after each lapping, and replaced the 4-point probe measurement. The technique seems to be potentially very valuable in diffusion work, but cannot be used much until the appropriate calibration curves are plotted. The results are shown in Fig. 7.9. At concentrations of atoms below about 10^{18} cm^{-3}, the two curves coincide, but above this concentration, the electron level saturates.

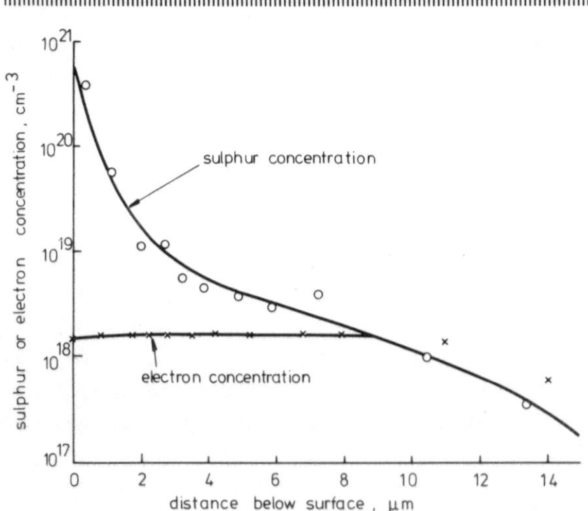

Fig. 7.9 Comparison of the electron concentration and the total sulphur concentration in a sample of GaAs diffused for 3 days at 900°C

Reproduced from Young, A. B., and Pearson, G. L.: *J. Phys. & Chem. Solids*, 1970, **31**, p. 517

* Young, A. B. Y., and Pearson, G. L., *J. Phys. & Chem. Solids*, 1970, **31**, p. 517

7.4 Diffusion coefficients from profiles

Let us suppose that a diffusion experiment has been satisfactorily carried out and the profile has been plotted using one or other of the methods outlined above. The next step is to use the profile to compute the diffusion coefficient. Take first the case for which the coefficient has a constant value. There are two common sets of diffusion conditions which must be considered. The first is the thin-film condition in which a quantity α of the solute is plated on to unit area of the surface of the sample. Diffusion takes place with the restriction that the total amount of material in the semiconductor is always α cm^{-2}. The calculation was described in Chapter 2, and the solution for the diffusion profile is

$$C(x) = \frac{\alpha}{(\pi Dt)^{1/2}} \exp\left(\frac{-x^2}{4Dt}\right)$$

A graph of ln C against x^2 should yield a straight line, from which D can be found.

The second common set of experimental conditions (and the one most often encountered in semiconductor work) is that of the infinite source. The semiconductor is brought into contact with an external phase containing the solute. The concentration at the surface immediately achieves the solubility C_0, and remains at this value throughout the experiment. This problem was also treated in Chapter 2, and the reader might remember that it was approached somewhat obliquely. The problem actually solved was one in which two semi-infinite rods are joined at $x = 0$. The bar stretching from $x = -\infty$ to $x = 0$ has a uniform concentration C' of the solute, and the other contains none. The concentration of solute at $x = 0$ immediately goes to $\frac{1}{2}C'$ which is equal to C_0, and remains there for the whole experiment. From the point of view of the bar covering $x = 0$ to $x = \infty$ this is the same condition as that of the infinite source. The two cases are therefore the same, and the solutions are mathematically identical. The solution is

$$C(x) = C_0 \operatorname{erfc} \frac{x}{2(Dt)^{1/2}} \tag{7.19}$$

The solution is for the 2-bar problem for all x, and for the infinite source problem for positive values of x.

If the error function is plotted on probability paper, it comes out as a straight line. Fig. 7.10 is a plot of $C/C' \times 100$ (or $C/C_0 \times 50$) against $x/t^{1/2}$ for eqn. 7.19 The factor of 100 is needed because probability paper is usually printed with the ordinate covering 0 to 100 rather than 0 to 1. For a diffusion carried out under infinite source conditions, therefore, one plots the experimental profile points on probability paper, as $C/C_0 \times 50$ against $x/t^{1/2}$. If a straight line is obtained, then it can be assumed that the diffusion coefficient is a constant. Having established that the profile is an error function complement, it remains to calculate the diffusion coefficient. This can be done using the following argument. From tables, we have erfc(1) = 0·157. From eqn. 7.19, when $C/C_0 = 0{\cdot}157$, $x/2(Dt)^{1/2} = 1$. The procedure

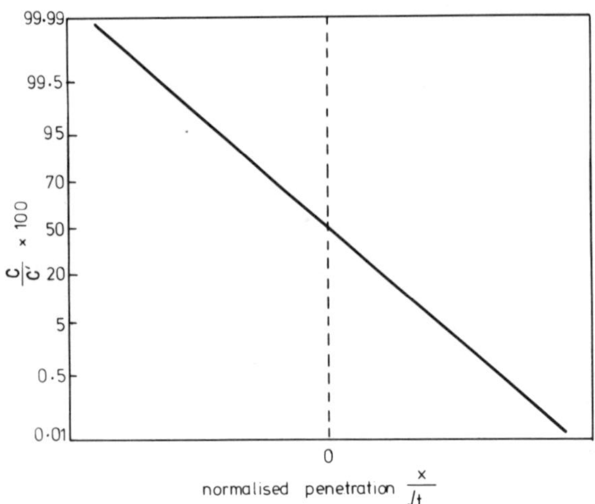

Fig. 7.10 Error function complement curve plotted on probability paper

is therefore to obtain from the profile the depth at which C/C_0 has the value 0·157. Call this x'. Then $x' = 2(Dt)^{1/2}$, where t is the duration of the diffusion experiment.

If a straight line is not obtained on the probability plot, the diffusion coefficient is not a constant, and life becomes very hard for the diffusion scientist. Under certain circumstances, however, it is still possible to calculate the coefficient from the profile. These circumstances which, fortunately, are quite common, are the following: (a) the diffusion coefficient is a function only of concentration, i.e. $D = D(C)$ (b) The initial conditions can be described in terms of the variable $\eta = x/t^{1/2}$. These two assumptions form the basis of the Boltzman-Matano technique for analysing diffusion profiles. When they are both obeyed, the concentration C can be shown to depend on the single variable η rather than the two variables x, t.

It is convenient to start once again by considering two semi-infinite bars, because we are then looking at the whole solution, from $x = -\infty$ to $x = \infty$. The infinite source case can then be taken as the positive half of the solution. The boundary conditions can be written

$$C = C' (= 2C_0) \text{ for } x < 0 \text{ at } t = 0$$

$$C = 0 \qquad \text{for } x > 0 \text{ at } t = 0$$

and these can be expressed in terms of η

$$C = C' \quad \eta = -\infty$$

$$C = 0 \quad \eta = \infty$$

So here the assumption concerning the boundary condition is valid. Now consider Fick's second law for a concentration-dependent diffusion coefficient

$$\frac{\partial C}{\partial t} = \frac{\partial}{\partial x}\left\{ D(C)\frac{\partial C}{\partial x} \right\} \tag{7.20}$$

Introducing the variable η, we have

$$\frac{\partial C}{\partial x} = \frac{\partial C}{\partial \eta}\frac{\partial \eta}{\partial x} = \frac{1}{t^{1/2}}\frac{dC}{d\eta}$$

and, similarly

$$\frac{\partial C}{\partial t} = -\frac{x}{2t^{3/2}}\frac{dC}{d\eta}$$

hence

$$\frac{\partial}{\partial x}\left(D\frac{\partial C}{\partial x}\right) = \frac{\partial}{\partial x}\left(\frac{D}{t^{1/2}}\frac{dC}{d\eta}\right) = \frac{1}{t}\frac{d}{d\eta}\left(D\frac{dC}{d\eta}\right)$$

Substituting into eqn. 7.20

$$-\tfrac{1}{2}\eta\frac{dC}{d\eta} = \frac{d}{d\eta}\left(D\frac{dC}{d\eta}\right) \tag{7.21}$$

so that now both the differential equation and the boundary conditions are expressed in terms of one variable.

Eqn. 7.21 can be integrated with respect to η between $C = 0$ and $C = C_1$, where C_1 is any specific value of C, i.e. $0 < C_1 < C'$:

$$-\tfrac{1}{2}\int_0^{C_1} \eta\, dC = \left[D\frac{dC}{d\eta}\right]_{C=0}^{C=C_1} \tag{7.22}$$

The profile under discussion will have been plotted for some specific time, so t can be treated as a constant when x and t are put back in place of η

$$-\tfrac{1}{2}\int_0^{C_1} x\, dC = Dt\left[\frac{dC}{dx}\right]_{C=0}^{C=C_1} = Dt\left(\frac{dC}{dx}\right)_{C=C_1} \tag{7.23}$$

where the last equality comes from the fact that at $C = 0$, $dC/dx = 0$. Since dC/dx is also zero at $C = C'$, we have from eqn. 7.23 that

$$\int_0^{C'} x\, dC = 0 \tag{7.24}$$

It follows that for the boundary conditions to be satisfied, the origin from which x is measured must obey eqn. 7.24, i.e. eqn 7.24 defines $x = 0$.

From eqn. 7.23, an equation is found for the diffusion coefficient at $C = C_1$

$$D(C_1) = -\frac{1}{2t}\left(\frac{dx}{dC}\right)_{C_1} \int_0^{C_1} x\,dC \qquad (7.25)$$

All of the quantities on the right-hand side of eqn. 7.25 are available from the experiment. The diffusion time t is known, and the other two quantities are the inverse of the slope at C_1 and the area under the graph.

The procedure is therefore as follows. The experimental profile is plotted on the largest piece of (linear) graph paper available, (Fig 7.11). The zero for the x axis is found using eqn. 7.24. In practical terms this means making the two hatched areas in Fig. 7.11a equal. The value of the diffusion coefficient corresponding to the concentration C_1 is then found by measuring the slope of the profile, and also the shaded area in Fig. 7.11b. Areas can be measured either by counting squares on the paper or by using a planimeter. The values are substituted in eqn. 7.25 and $D(C_1)$ calculated. The process is repeated for other values of C_1, and the function $D(C)$ emerges.

When the experiment is 'infinite source' rather than two semi-infinite bars, we have only to solve for positive x. This means that the $x = 0$ position is not fixed by an equation like eqn. 7.24: it is fixed by the experiment at the surface of the specimen. It is important to remember, however, that one is dealing with only half the solution, and not to get C', the largest concentration measured in the 2-bar experiment, mixed up with C_0, the largest measured in the semiconductor experiment. In fact, $C_0 = ½C'$.

Concentration-dependent diffusion coefficients are quite common in semiconductor systems, so the Boltzman-Matano technique is used a good deal in analysing experimental diffusion profiles. Earlier in the chapter, the isoconcentration technique for determining the function $D(C)$ was described. It involves a lot of experimentation and is obviously more trouble than the graphical technique, which

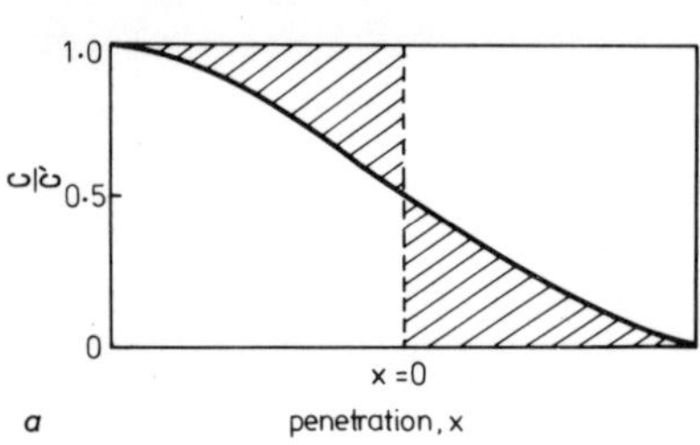

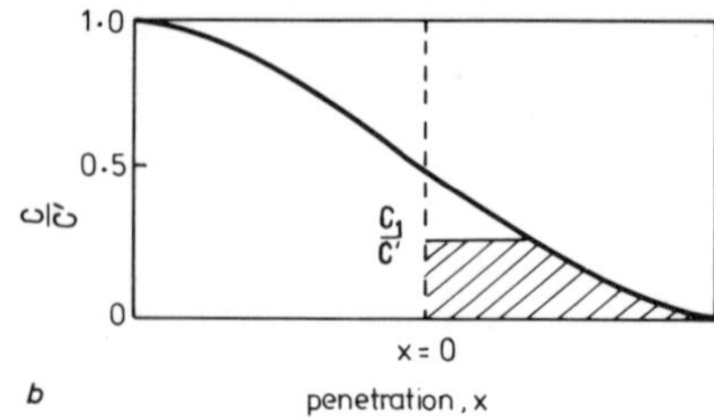

Fig. 7.11 Boltzman-Matano technique

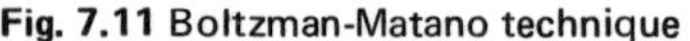

gets all the information from a single profile. The isoconcentration technique is more accurate, however, and is based on fewer assumptions. The assumption at the core of the Boltzman-Matano technique is that the two variables x and t can be replaced by the single variable $x/t^{1/2}$. This is clearly correct when D is a constant, since in eqn. 7.19 they only appear in this form. The Boltzman-Matano analysis is only used, however, in cases when D is a variable, and strictly the assumption should be checked experimentally for each system. This can be done by carrying out a series of diffusions in which the conditions are identical in every respect except time. The experiments should cover as large a range of time as possible. The resulting profiles, when plotted with $x/t^{1/2}$ as the abscissa, should be coincident. The experiment has not been carried out for many semiconductor systems, and where it has been done, the analysis has sometimes been shown to be invalid. The Boltzman-Matano method of profile analysis must therefore be used with caution: for those cases to which it is valid, it is an important and extremely useful technique.

Chapter 8

Diffusion and dislocations

So far almost all of the discussion about defects in crystals has been restricted to point defects. In this chapter, we look at other types of defects and consider their interaction with the diffusion process. Primarily, we shall be concerned with dislocations, but a little will be said about precipitates. It will be shown that the diffusion process can cause both dislocations and precipitates to form in a crystal and that dislocations can themselves have an effect on diffusion. Hence an interesting 'feedback' situation can arise in which diffusion creates dislocations which then have an effect on subsequent diffusion.

8.1 Edge and screw dislocations

In general, plastic deformation of single crystals takes place by planes of atoms slipping over one another, rather like a pack of playing cards. It is possible to calculate, for a perfect crystal, the stress required to bring the deformation about: when this is done, an answer is obtained which is some orders of magnitude greater than the stress required in practise. In other words, crystals are weaker than a simple theory would predict. The existence of dislocations was originally postulated to explain this weakness, and there are now many experimental techniques whereby they can be directly observed.

There are two principal types of dislocation; the edge and the screw. A diagram of an edge dislocation is given in Fig. 8.1a: it is often described as a region of perfect crystal into which an extra halfplane has been forced. The diagram is for a simple cubic lattice and does not refer to any specific semiconductor, since all semiconductors have a more complicated lattice structure than this. This diagram (Fig. 8.1a) is adequate to illustrate all of the points to be made in this chapter, however, and will be used for the sake of simplicity. The dislocation line is at right angles to the plane of the diagram, at the bottom of the extra halfplane. The dislocation can move through the crystal very easily. In Fig. 8.1a it is centred over position A. If the shearing stress indicated in the diagram is applied, however, it can

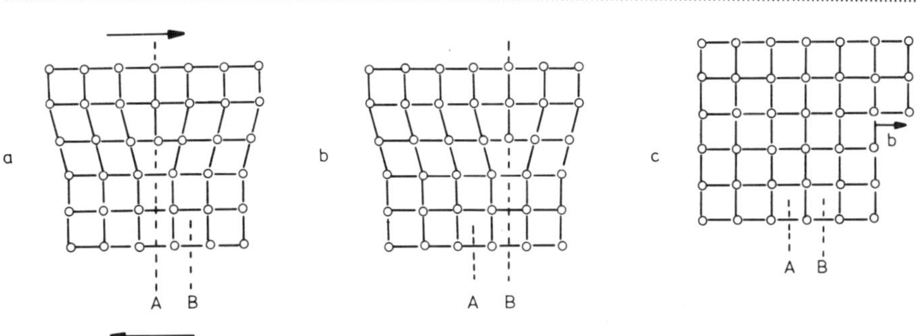

Fig. 8.1 Movement of an edge dislocation through a region of crystal

move to the right so that it is centred over position B, as in Fig. 8.1b. A comparison of the two diagrams shows that only a very small movement is required on the part of each of the atoms, so the dislocation can move under the action of a fairly low stress. If the stress is applied continuously, the dislocation line continues to move to the right until it effectively falls out of the crystal. When this happens, the top of the crystal has slipped an amount b relative to the bottom, as shown in Fig. 8.1c. The amount of slip associated with a dislocation (b in this case) is called the Burgers' vector.

The other type is called a screw dislocation and is shown in Fig. 8.2. Once again the dislocation line runs at right angles to the plane of the diagram, but this time the Burgers' vector is parallel to the line rather than perpendicular. This means that there is no 'extra halfplane' associated with a screw dislocation. The normal motion of the line is again at right angles to itself. Most dislocation lines are neither pure edge nor pure screw. The dislocation loop shown in Fig. 8.3, for example, has the Burgers' vector in the same plane as the loop, but is pure edge only at points A and C, and pure screw at B and D. At all other points on the loop it is a mixture of edge and screw.

Most crystals have a number of possible slip systems that can operate, so that when conditions are such as to allow the dislocation lines to move, they can change the crystal to any desired shape, providing the right stresses are applied. Under these conditions the crystal is said to be plastic. At low temperatures, the dislocations are 'frozen' in the crystal, and cannot move. The application of a stress then produces only a small amount of elastic deformation, i.e. on removal of the stress, the sample returns to its original shape. If a large stress is applied, the crystal breaks in a brittle manner owing to cracks propagating through it. Most pure metals are plastic at room temperature. Semiconductors are brittle at room temperature, but diffusion temperatures are usually high enough to take them into the plastic region.

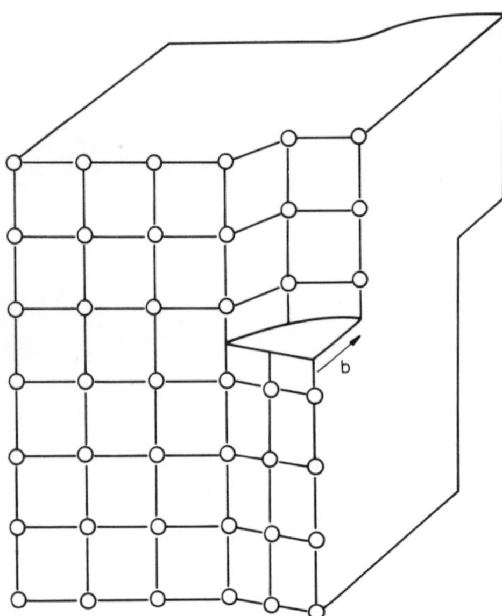

Fig. 8.2 Screw dislocation

From what has been written so far, it might appear that the best way to obtain a dislocation-free crystal is to apply a stress so that all of the dislocations run out of the crystal, as in Fig. 8.1. This is not so. Dislocation sources exist in the crystal which, under conditions of plastic deformation, throw out large numbers of dislocations when the stress is applied. When a crystal undergoes plastic flow,

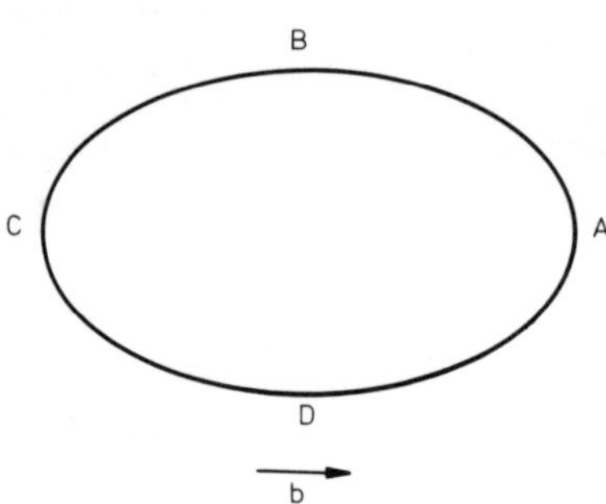

Fig. 8.3 Dislocation loop

therefore, the number of dislocations in the specimen increases enormously. In metals the increase causes the dislocations to become tangled up with each other so that eventually they can be moved only by the application of an increasingly larger stress. This is the familiar phenomenon of work hardening. As well as being activated by an externally applied stress, dislocation sources can also be started by internal stresses. The dislocations that are produced then relieve the internal stresses. This is the means whereby dislocations are produced during diffusion.

Density of dislocations is expressed in terms of the number of dislocation lines crossing unit area in the crystal. It is possible to produce semiconductor crystals, notably of silicon, with virtually no dislocations at all, but it takes very careful handling to keep them in that condition. Many semiconductor crystals are produced with dislocation densities of the order of 100 cm^{-2}. If the crystals are subjected to stress, by abrading for instance, then the density increases by a large factor, to perhaps 10^5 cm^{-2}. In metals, much higher dislocation densities are commonly found. A piece of metal that has been worked in some way might have a density in the region of 10^{10} cm^{-2}.

8.2 Generation of dislocations by diffusion

When we diffuse an element into a crystal, we are introducing solute atoms into a solvent lattice in which the atoms are of a slightly different size. The solute atoms do not fit properly, therefore, and stresses arise in the crystal. The situation is portrayed in Fig. 8.4, which represents an end view of a crystal that has been diffused in the x direction. The solute atoms are smaller than the atoms of the crystal. The layer of atoms a (Fig. 8.4) contains the largest number of solute atoms, so it wants to be smaller in its y and z dimensions than the layers beneath. Because the layer is joined to the layers underneath, however, it cannot contract as much as it wants to: it is therefore under tension in the y and z directions. Call these stresses

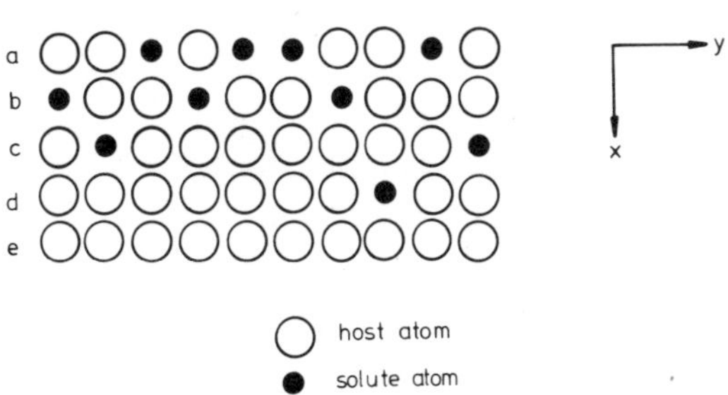

Fig. 8.4 Crystal containing a concentration gradient of solute atoms which are smaller than the host atoms

σ_y and σ_z. In an isotropic crystal they will be equal. Note that there is no equivalent stress in the x direction because all layers in x–z planes contain the same number of solute atoms, on the average.

The bottom layer, e (Fig. 8.4), contains no solute atoms. The effect of the contracted layers to which it is joined is to cause the layer to be slightly smaller than the normal size for the pure crystal, i.e. the layer is under a compressive stress in the y and z directions. The planes therefore change from a state of tension at the top of the crystal, where the solute concentration is a maximum, to a state of compression at the bottom. Somewhere in between there must be a plane for which σ_y and σ_z are zero.

The problem was considered by Prussin (1961) who realised that, mathematically speaking, it was the same as another problem that had already been solved. If a 1-dimensional temperature gradient is imposed on an elastic crystal, then adjacent planes of the crystal are at different temperatures. They therefore want to expand by different amounts, and stresses arise just as for diffusion. The problem is solved in the book by Timoshenko and Goodier (1951). The only difference between the two situations is that in the first the crystal wants to change its dimensions in proportion to the solute concentration in the layer, whereas in the second it wants to change proportionately to temperature. Prussin (1961)* considered a semiconductor slice, thickness $2a$, as shown in Fig. 8.5. Diffusion is assumed to take place from the surfaces at $x = a$ and $x = -a$. This is a distinct advantage from the mathematical point of view: if diffusion occurred from only one of the surfaces, the slice would buckle and it would be necessary to include an extra term to allow for this. Replacing temperature by concentration in Timoshenko's (Timoshenko and Goodier, 1951)† formula gives an expression for the stress in the specimen

$$\sigma_y = \sigma_z = \frac{\beta Y}{3(1-\nu)}\left(C - \frac{1}{2a}\int_{-a}^{a} C dx\right) \tag{8.1}$$

where C is the concentration of solute and β is the solute lattice contraction coefficient, i.e. the decrease in volume of the crystal per added solute atom. Y is Young's modulus and ν is Poisson's ratio. It has been assumed that the crystal is isotropic, so that the stress in the y direction is the same as that in the z direction. A positive value of stress in eqn. 8.1 indicates that the lattice planes are in tension at that value of x. The integral term in eqn. 8.1 is the average amount of solute in the slice, C_{av} say, so the equation can be written more simply

$$\sigma_y = \frac{\beta Y}{3(1-\nu)}(C - C_{av}) \tag{8.2}$$

Eqn. 8.2 confirms a conclusion reached earlier, namely that one y–z plane in the

* Prussin, S., *J. Appl. Phys.*, 1961, **32**, p. 1876

† Timoshenko, S., and Goodier, J. N., 'Theory of Elasticity', 2nd Edn., McGraw-Hill, 1951.

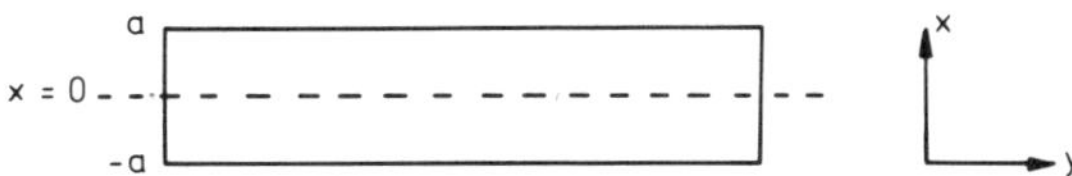

Fig. 8.5 Semiconductor slice, diffused from surfaces at $x = \pm a$

slice has zero stress: this is at the value of x at which the concentration of solute C is equal to the average concentration C_{av}. On one side of the plane the crystal is in tension, and on the other side it is in compression. The maximum stress arises where C has its maximum value and C_{av} is a minimum. For both constant-source and infinite-source conditions, this is right at the start of the diffusion, when the surface boundary condition can be written

$$C = C_0 \qquad x = 0, t = 0$$

$$C = 0 \qquad x > 0, t = 0$$

The second condition gives $C_{av} = 0$ at $t = 0$, so eqn. 8.2 becomes

$$(\sigma_y)_{max} = \frac{\beta Y C_0}{3(1 - \nu)} \tag{8.3}$$

The physical constants appearing in eqn. 8.3 are known for many materials, so $(\sigma_y)_{max}$ can be calculated for a typical surface concentration. The minimum stress needed to activate dislocation sources is also known for many materials. Call this σ_d, Prussin (1961) took as an example the diffusion of phosphorus in silicon. He compared σ_d with $(\sigma_y)_{max}$ and found that the stress introduced by the phosphorus was more than enough to cause the generation of dislocations. It seems likely that the same is true for other semiconductor systems.

Inspection of eqn 8.2 shows that $\sigma_y (= \sigma_z)$ and C differ only by a multiplying factor and a constant term, i.e. the two functions are essentially of the same form. Thus, using suitable units and different origins for the two quantities, a single curve can be used to represent both as a function of diffusion depth. This is done in Fig. 8.6. The graph shows the profile (and stress distribution) for the diffused layer beneath the surface at $x = a$. Eqn. 8.2 was derived on the assumption that no stress relief takes place, so Fig. 8.6 represents the stress distribution, if there are no dislocations generated. If $\sigma_d < \sigma_y$ at any point, then dislocations are generated and the stress drops to σ_d.

Consider what happens to the stresses as the diffusion experiment proceeds. Assume the solute atoms are smaller than those of the solvent so that the surface stress is tensile. At the beginning of the experiment the stress at the surface is $(\sigma_y)_{max}$. If this is larger than σ_d, dislocations are formed and the stress falls to σ_d. As the diffusion profile penetrates further into the specimen, the layer immediately

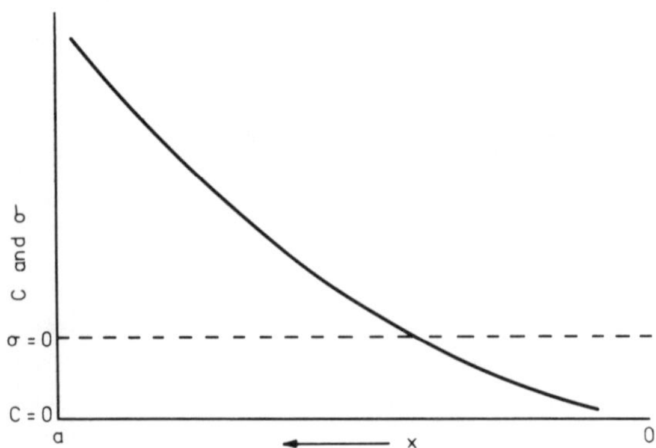

Fig. 8.6 Diffusion profile and stress distribution for the sample of Fig. 8.5. No generation of dislocations

below the first also achieves a tensile stress greater than σ_d and more dislocations are formed. The stresses close to the surface are therefore smaller than those shown in Fig. 8.6, and this has the effect of pushing the layer of zero stress deeper into the crystal. The situation might therefore be summarised as follows: as diffusion proceeds, a 'dislocation front' moves into the crystal, the region behind the front contains dislocations and the stress within it is uniform at σ_d. The region ahead of the front contains no dislocations and follows the stress curve of Fig. 8.6. The plane of zero stress is deeper inside the crystal than indicated in Fig. 8.6. The new stress distribution is shown in Fig. 8.7, in which the edge of the dislocation front is at $x = b$.

It is helpful to consider how the dislocations can physically relieve stresses caused by solute atoms. Consider an element of the slice, with sides of unit length, and thickness dx, as shown in Fig. 8.8. Suppose that the surface at $(x + dx)$ is nearer the top surface of the slice, so that it contains more solute atoms than the surface at x. The concentration of solute at x is C, and that at $(x + dx)$ is $\{C + (\partial C/\partial x)dx\}$. Since the solute atoms are smaller than the host atoms here, the surface at $(x + dx)$ would, left to itself, be physically smaller than the surface at x. It cannot manage this because it is attached to the adjacent planes. Now introduce a number of edge dislocations with the dislocation lines in the y–z plane and the extra halfplanes vertical and on the $+x$ side of the lines. If there are more dislocation lines in the $(x + dx)$ surface than in the x surface, then more halfplanes finish on the former than the latter. This means that there are actually more lines of atoms on the $(x + dx)$ surface than on the other surface. The increased number of atoms on the $(x + dx)$ surface relative to the crystal planes beneath, has the effect of making that surface larger. This compensates for the extra solute atoms, which

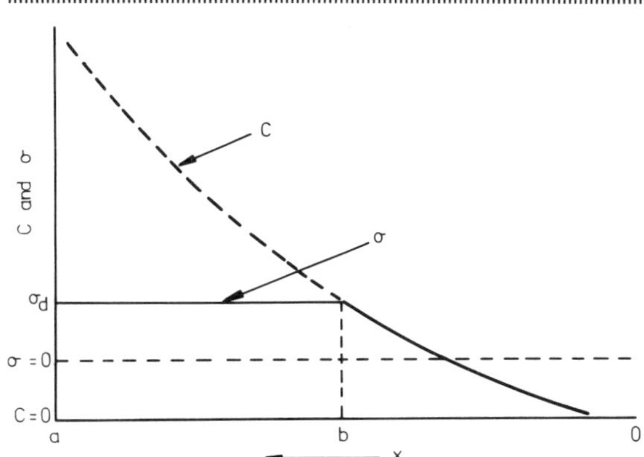

Fig. 8.7 Diffusion profile and stress distribution for the sample of Fig. 8.5, after dislocations have been allowed to generate

tend to make it smaller, and makes adjacent planes more nearly equal in size. Since the stresses are due to the planes trying to achieve different sizes, most of the stresses are relieved.

The point becomes more clear if we calculate the equilibrium number of dislocations. Consider the contraction of the element of Fig. 8.8 in the z direction, Δz. The lattice contraction between x and $(x + dx)$ is given by

$$\Delta z = \tfrac{1}{3}\,\beta\left(C + \frac{\partial C}{\partial x}\,dx\right) - \tfrac{1}{3}\,\beta C$$

$$= \tfrac{1}{3}\,\beta\left(\frac{\partial C}{\partial x}\right)dx \qquad (8.4)$$

where the bulk contraction coefficient β has been used rather than the linear coefficient, and we have used the fact that for small contractions the bulk strain is

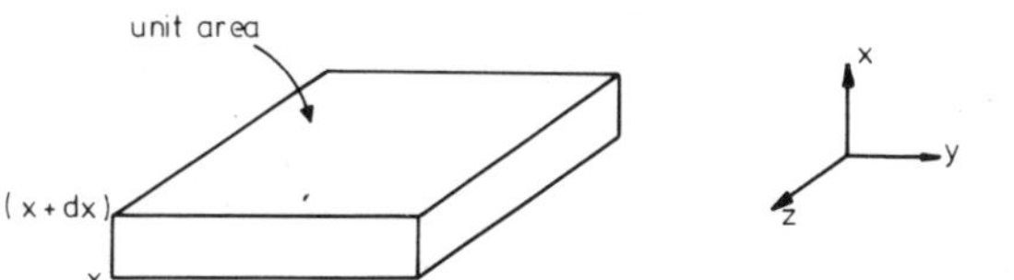

Fig. 8.8 Element of semiconductor slice

three times the linear strain. To relieve the stress caused by eqn. 8.4, an equivalent expansion can be brought about by introducing dislocations. Now only edge dislocations are of any use to us here, since it is the extra halfplane that is employed. Furthermore we are only interested in the z direction at present. Most dislocations are mixed edge and screw, so let λ be the component in the z direction of the edge part of the Burgers' vector. This will be the contribution to the extension per dislocation. If the density of dislocations is ρ, this makes the total extension in the z direction

$$\Delta' z = \lambda \rho \, dx \tag{8.5}$$

since the volume of the slice is dx. Now if the relative contraction due to the solute is to be compensated by the extension due to the dislocations, $\Delta z = \Delta' z$, and we have an expression for the dislocation density

$$\rho = \frac{\beta}{3\lambda} \cdot \frac{\partial C}{\partial x} \qquad b < |x| < a \tag{8.6}$$

$$\rho = 0 \qquad 0 < |x| < a \tag{8.7}$$

The modulus signs are necessary because we diffused into the sample of Fig. 8.5 from both surfaces. Note that by adding the dislocations we have not removed all of the strain from the regions $b < |x| < a$. We have merely removed the differential strain so that the stress is uniform in the region, as shown in Fig. 8.7.

The total number of dislocations N can now be calculated by integrating eqn. 8.6 between $x = a$ and $x = b$. This gives

$$\mathrm{N} = \frac{\beta}{3\lambda} \left\{ C_0 - C(b) \right\} \tag{8.8}$$

where $C(b)$ is the concentration of solute at $x = b$. It often happens that $C_0 >> C(b)$, in which case the total number of dislocations is proportional to the surface concentration. This result has an interesting consequence. It implies that the number of dislocations does not increase with time during a diffusion because, in general, the surface concentration of solute does not increase with time. The point can be made more rigorously for infinite source diffusion. Substituting the solute concentration profile

$$C = C_0 \operatorname{erfc} \frac{(a - x)}{2(Dt)^{1/2}} \tag{8.9}$$

into the dislocation density expression, eqn. 8.6 yields

$$\rho = \frac{\beta}{3\lambda} \frac{C_0}{(\pi D t)^{1/2}} \exp \left\{ - \frac{(a - x)^2}{4Dt} \right\} \qquad b < |x| < a \tag{8.10}$$

So the dislocation distribution is Gaussian. The integral of eqn. 8.10 is a constant, as shown in Chapter 2, so the total number of dislocations remains the same throughout the experiment. This ties in with the fact that the surface concentration is constant for infinite source conditions. (There is an approximation involved in this statement: the integral of eqn. 8.10 is strictly constant only if the integration is taken to infinity. In fact the distribution is cut off at $x = b$, after which the dislocation density is zero. The approximation is similar to the one made concerning eqn. 8.8, that $C_0 >> C(b)$.)

Eqn. 8.10 shows that at the beginning of the diffusion process a large number of dislocations is created at the surface. With increasing diffusion time, the stress imposed by lattice contraction gets smaller at the surface. The transport of solute atoms into the crystal increases the stress there, and dislocations flow from the surface into the crystal. The density of dislocations at the surface decreases throughout the experiment.

For constant source diffusion the surface concentration of solute falls with time (see Chapter 2). The argument given above suggests that the number of dislocations in the specimen must therefore decrease with time of diffusion. The excess dislocations can escape by slipping out of the free surfaces of the slice. Substituting the appropriate diffusion profile for constant source conditions into eqn. 8.6 shows that here the maximum dislocation density occurs not at the surface, but at a depth $(2Dt)^{1/2}$ below it.

Essentially the same arguments can be applied for contraction in the y direction as have been used above for that in the z direction. Finally it should be noted that throughout this section the example has been taken of a solute atom which is smaller than the atoms of the crystal into which it is diffusing. Similar theory can be applied for a solute atom which is larger.

8.3 Observation of diffusion-induced dislocations

The simplest way to show up the dislocations is to use a dislocation etch. After diffusion the slice is sectioned at right angles to the faces of the slice, i.e. at right angles to the surfaces $x = \pm a$ in Fig. 8.5. The cross-section thus revealed is treated with the etch, which develops a pit at any point where a dislocation intersects the surface. The experiment was carried out by Prussin to test his theory. He found a layer of etch pits close to the free surfaces at $x = \pm a$, but no pits in the centre of the specimen, in agreement with the prediction of the theory. He also found that for a specimen diffused using infinite-source conditions, the largest concentration of etch pits was just beneath a surface. For a specimen diffused using constant-source conditions, however, the maximum density was a little further towards the middle of the slice. This behaviour is also predicted in the preceding theory.

A series of experiments using transmission electron microscopy was carried out by Joshi and Wilhelm (1965)*. This technique has the advantage that individual dislocation lines can be seen and identified, but has the drawback that it is necessary to reduce the specimen to a very small thickness so that it is transparent to the electron beam. Initially phosphorus was diffused into silicon. The phosphorus atom is smaller than silicon, so it tends to contract the lattice. The electron micrographs indicated a very large dislocation density, compared to undiffused material, in the region of diffusion. Silicon which had been homogeneously doped with phosphorus during growth, to a high level, was also looked at with the microscope. The dislocation density for this material was low. This result suggests that it is not a high concentration of solute which gives rise to dislocations, but a concentration gradient. Again the result agrees with theory (eqn. 8.6). Joshi and Wilhelm (1965)* also carried out experiments using arsenic as the solute atom. The arsenic atom is almost the same size as silicon, so it would be expected that it would give rise to very few extra dislocations. Once again the prediction was confirmed by experiment.

The very good agreement with theory of the experiments of Prussin (1961) and of Joshi and Wilhelm (1965) strongly suggests that they were indeed observing dislocations caused by a concentration gradient of solute atoms which did not fit properly into the lattice. A number of workers, however, have observed diffusion-induced dislocations which lie much further below the surface than would be predicted by Prussin's theory. The deep-lying dislocations often appear when an oxide layer is used on the semiconductor slice and it seems quite likely that the layer may contribute towards their formation. The contribution could take the form of extra stresses being present in the semiconductor because of the presence of oxide. Alternatively the oxide may promote the growth of precipitates, by rejecting solute as it grows into the semiconductor. The effect was studied by Fairfield and Schwuttke (1968)†, again for phosphorus diffusing into silicon, using the technique of x-ray topography. They were able to divide the dislocations they observed into two types; the shallower ones which corresponded to those predicted by Prussin's theory, and the deeper ones. They found that this second category of dislocations could be made larger or smaller by varying the experimental conditions. The deep dislocations were most numerous in experiments in which high surface concentrations and oxidising ambients were used.

Fairfield and Schwuttke also observed precipitates of phosphorus in their work. The formation of precipitates during diffusion has been reported by several research workers, for a variety of semiconductor systems. It seems likely that these will give rise to stresses within the slice which are not allowed for in the simple theory. The observation of precipitates also has relevance to the electrical properties of the doped semiconductor. It was pointed out in Chapter 7 that diffusion profiles plotted using electrical techniques sometimes do not agree with profiles arrived at

* Joshi, M. L., and Wilhelm, F., *J. Electrochem. Soc.*, 1965, **112**, p. 185

† Fairfield, J. M., and Schwuttke, G. H., *J. Electrochem. Soc.*, 1968, **115**, p. 415

using radiotracer methods. Precipitation is one possible reason for this. If a doping atom occurs in the semiconductor as part of a precipitate, rather than simply appearing on a lattice site, it is most unlikely that it will have the expected electrical characteristics. It is quite likely that it will not be electrically active at all.

In Chapter 7 a technique was described for finding p–n junction depths using a chemical etch. The diffused specimen is sectioned perpendicularly to the surface. It was noted that while the technique usually works well, it sometimes works rather too well, and more than one line is obtained. It is then necessary to decide which one represents the junction. Prussin's theory throws some light on this. A series of etching experiments was carried out by Schwuttke and Rupprecht (1966)* on GaAs diffused with zinc. Three etching lines were observed. They also looked at the specimens using x-ray topography and this enabled them to identify the lines. The line nearest the surface proved to be the 'dislocation front', i.e. it marked the limit of the region containing a high concentration of dislocations. The second line gave the p–n junction. The third line, furthest from the surface, was the edge of the region of high elastic strain due to the presence of the solute.

8.4 Effect of dislocations on devices

The discussion of the interaction of diffusion with dislocations is by no means purely academic. Many semiconductor devices are made by diffusion and there is a good deal of experimental evidence to show that the presence of dislocations can affect the performance of a device to a marked extent. The diffusion process is not responsible for all the dislocations that find their way into devices, of course. Other processes such as mechanical polishing, oxidation and even handling with tweezers can all introduce mechanical defects into semiconductors. However, a high concentration of dislocations can be caused by diffusion and it is worth considering what effects they can have on electrical properties.

Most theoretical approaches to the electrical effects of dislocations are based on the 'dangling bond' concept (Read, 1954)†. Consider the edge dislocation shown in Fig. 8.1. The atoms in the part of the lattice which is perfect are bonded to four other atoms in the plane of the diagram. The atom which is actually on the dislocation line, i.e. the one at the bottom of the halfplane, is bonded only to three other atoms, however. There is therefore an unpaired electron, or dangling bond, associated with the atom, and with every other atom on the dislocation line. In Fig. 8.1, a dislocation in a simple cubic lattice is shown but it is generally true that associated with an edge dislocation line is a line of dangling bonds. The spare electron associated with the bond might be able to break away and become a free electron in the conduction band. Here it would be behaving as a donor.

* Schwuttke, G. H., and Rupprecht, H., *J. Appl. Phys.*, 1966, **37**, p. 167

† Read, W. T., *Phil. Mag.*, 1954, **45**, p. 775

Alternatively it might attract a second electron from the valence band and become paired with it. Then it would be an acceptor. It is not immediately obvious which of the possibilities is the more likely, but experiments have usually demonstrated acceptor behaviour on the part of edge dislocations. Either way, the effect of the defect is to introduce extra energy levels into the forbidden gap of the semiconductor.

Suppose a dislocation line acts as a line of acceptors. If a large proportion of its atoms accepts electrons, the dislocation acts as a negatively charged line in the crystal. The line repels conduction electrons and attracts holes, so that under suitable conditions a cylindrical space-charge region forms around the dislocation. In principle, this situation gives plenty of scope for device properties to be affected. The presence of extra levels in the forbidden gap gives rise to extra recombination paths for minority carriers, and can severely reduce the minority carrier lifetime. The line of charge associated with a dislocation produces extra scattering of carriers and can significantly affect electron and hole mobilities.

In addition to these essentially electronic effects there is also the fact that doping atoms tend to congregate about dislocations. The effect occurs because there is more space at a dislocation to accommodate atoms which do not fit well into the lattice (Fig. 8.1). The solute concentration close to a dislocation is therefore likely to be different to that in the rest of the lattice. It follows that the electrical properties in that region might also be different.

A number of workers have related device performance to the dislocation content of the semiconductor. Queisser and Goetzberger (1963)*, for instance, presented some striking evidence linking microplasma breakdown in silicon p–n junctions with dislocations. Reverse-bias breakdown in these junctions can occur by a junction giving way at a number of discrete points. At these points, microplasmas occur, and a very high current density flows across the junction. The points give out light and can therefore be easily identified. Queisser and Goetzberger prepared silicon p–n junction diodes containing dislocations which crossed the junction. The positions of the dislocations were revealed using a metallographic technique, and the diodes were biased to reverse breakdown. The spots of light from the microplasmas were shown to correspond to the places on the junction where the dislocations crossed.

Similar results have been obtained for semiconductors other than silicon. Mozzi and Lavine (1970)† made a number of p–n junction diodes in InSb and studied the dislocation structure using x-ray topographic techniques. The diodes were prepared by diffusing zinc into n-type material. Two types of diffusion were carried out. In the first, the surface concentration was less than 10^{19} cm^{-3}, and in the second, it was greater than 10^{20} cm^{-3}. It was found that no diffusion-induced dislocations were formed in the first type and that the diodes had good electrical characteristics.

* Queisser, H. J., and Goetzberger, A., *Phil. Mag.*, 1963, 8, p. 1063

† Mozzi, R. L., and Lavine, J. M., *J. Appl. Phys.*, 1970, **41**, p. 280

The rectifying properties were excellent and a good photoresponse was obtained. The second type of diffusion gave a high dislocation concentration below the surface and very poor electrical characteristics.

Many more results could be quoted from the literature to establish that dislocations can have an important influence on device operation. While the dangling-bond theory can account for these results in a qualitative way, it does not yet appear to be sufficiently sophisticated to explain them in any detail.

8.5 Effect of dislocations on diffusion

Having considered the ways in which the diffusion process can give rise to dislocations, let us now turn the subject round and look at the ways in which dislocations can affect diffusion. The dislocations in question might be due to the diffusion that has already taken place, or to some other cause, such as mechanical polishing, etc.

The most straightforward way that a dislocation can influence diffusion is to act as a pipe of high diffusivity. Reference to the edge dislocation of Fig. 8.1 shows that immediately below the halfplane is a 'hole' of atomic dimensions. This hole extends along the length of the dislocation and can provide a path of easy diffusion for a solute atom. In general, it would appear that a rather high dislocation content is required before the overall diffusivity in the lattice is increased. Even if the transport rate along the pipe is high, the process still depends on the rate at which solute atoms leave at the far end of the pipe, and this will depend on the normal lattice diffusion coefficient.

Whether dislocation pipes are instrumental in increasing overall diffusion coefficient or not, it is certainly true that the diffusing solute atoms take advantage of the extra space close to a dislocation line and congregate there. This was first established by Dash (1956)*, who studied copper diffusing in silicon. Now silicon is transparent to infrared light. Dash used a technique in which infrared light was shone through a silicon slice, which was observed under a microscope fitted with an infrared image converter. After diffusing the silicon with copper, the specimen was etched in a dislocation etch and then viewed with the infrared microscope. The specimens proved to be transparent except for a number of dark lines. All of these lines started and finished at a free surface, and the points of intersection with the surfaces corresponded to dislocation etch pits. Dash came to the very reasonable conclusion that the diffusing copper atoms had migrated preferentially to the dislocation lines and 'decorated' them. Since the copper precipitates were not transparent to infrared, the dislocations became visible under the microscope. The technique of decoration is now used by many laboratories as a standard test for observing dislocation structure in semiconductors.

* Dash, W. C., *J. Appl. Phys.*, 1956, **27**, p. 1193

There is a second mechanism whereby dislocations can affect diffusion which is probably more important in practise. Diffusion mechanisms which are not purely interstitial all depend on the vacancy concentration in one way or another, so that any mechanism the crystal might have for changing the vacancy concentration is of importance. Simple substitutional mechanisms, for instance, give rise to a diffusion coefficient that is proportional to the vacancy concentration. In principle the equilibrium vacancy concentration is fixed by the conditions of the experiment, but diffusion is not an equilibrium process and one must always consider the possibility of the number of vacancies being 'wrong'. The point becomes even more important in the substitutional-interstitial mechanism, described in Chapter 6. This process actually consumes vacancies, so the crystal is continually faced with the problem of trying to replace them. There are two principal ways that a crystal can create vacancies during diffusion. Vacancies can diffuse into the crystal from a free surface, or they can be produced in the bulk of the crystal by the mechanism of dislocation climb.

Consider once again the edge dislocation of Fig. 8.1. The usual way for it to move is in the direction of the Burgers' vector, as shown in the Figure. Under certain circumstances the line can move in a direction perpendicular to both itsef and the Burgers' vector, i.e. up or down in Fig. 8.1, and this motion is called climb. The dislocation moves up by one lattice spacing if the bottom row of atoms of the extra halfplane diffuses away to occupy interstitial positions in the lattice. Similarly the line moves down by one space if atoms leave the lattice and join the bottom of the halfplane. This second process leaves vacancies behind in the lattice, and therefore changes the vacancy concentration. A screw dislocation can also climb by the acquisition of vacancies or atoms. Here the addition of the extra components causes the screw to form a helical dislocation, with the axis of the helix parallel to the Burgers' vector, i.e. parallel to the original direction of the screw dislocation.

Climb of dislocations during diffusion has been observed by Dash (1960)*, studying the diffusion of gold in silicon. It is likely that gold diffuses by a substitutional-interstitial mechanism, so the crystal has to find a means of making vacancies. Gold was diffused into silicon crystals containing a moderate density of dislocations. Copper was also diffused in, to decorate the dislocations and make them visible under the infrared microscope. The experiments showed that the effect of diffusing the gold was to introduce a number of helical dislocations, which were not otherwise present. As noted above, the presence of these dislocations indicates that climb must have taken place during diffusion.

It would be wrong to give the impression that dislocation climb usually involves the co-operation of the whole dislocation line. Most commonly, it happens because of the presence of a step, or 'jog' in an otherwise straight section of dislocation. These jogs are formed if two dislocation lines cut across each other. In Fig. 8.9 is shown the situation just after two pure screw dislocations have crossed. Each has

* Dash, W. C., *J. Appl. Phys.*, 1960, **31**, p. 2275

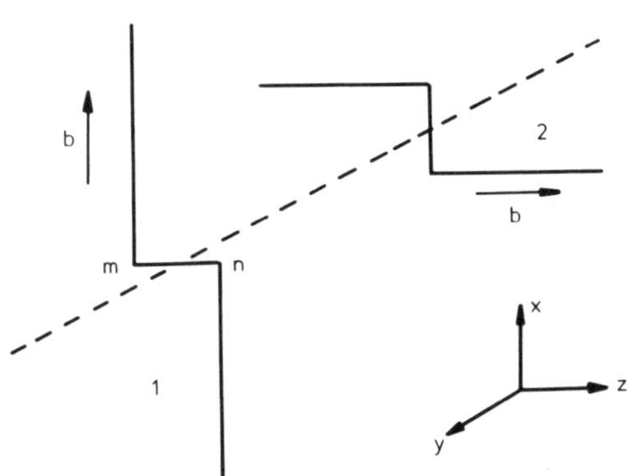

Fig. 8.9 Jogs produced in a pair of screw dislocations after they have intersected

left a jog in the other, equal to the respective Burgers' vectors. In dislocation 1, the new section, marked *mn* in the Figure, is a piece of edge dislocation. Its natural direction of movement is along the dislocation, i.e. along the x-axis. The rest of the dislocation wants to move along the y-axis, however. This is a direction which is perpendicular to both the Burgers' vector and *mn* and therefore represents a climbing motion for *mn*. It follows that as the dislocation moves along the y axis, it must leave behind it a trail of vacancies or of interstitials, depending on the direction of climb.

Phosphorus diffuses in an anomalous manner in silicon, and the generation of vacancies by dislocations has been suggested as a possible cause. The diffusion mechanism is substitutional and for low concentrations of phosphorus, a constant diffusion coefficient is obtained. Above a certain concentration, however, a larger diffusion coefficient is measured. It was noted by Parker (1967)* that for high concentrations of phosphorus, diffusion-induced dislocations are formed. According to Prussin's theory, the dislocation distribution within a sample changes as diffusion proceeds. This involves movement of the dislocations and, if they contain jogs, vacancies are emitted. The vacancy concentration of the crystal is therefore increased, and this shows up as an increase in the diffusion coefficient. The same explanation works equally well for the anomalous diffusion of boron in silicon.

8.6 Emitter-dip effect

Both in this chapter and in Chapter 4 we have considered ways in which defects can influence diffusion behaviour. Finally, we consider a phenomenon of

* Parker, T. J., *J. Appl. Phys.*, 1967, **38**, p. 3471

importance in transistor manufacture which is certainly due to the interaction of defects with the diffusion process, although it cannot yet be claimed that it is perfectly understood. A double-diffused *n–p–n* transistor is made using two diffusion stages. First an acceptor is diffused into an *n*-type slice, forming a *p–n* junction. The *p* region forms the base, and the *n* region the collector of the transistor. The emitter region is formed by diffusing in an *n*-type dopant for a small distance below the surface, forming the *n–p–n* structure. During the second diffusion, further diffusion of the *p*-type dopant takes place. When this process is carried out in silicon, the *p*-type dopant below the emitter diffuses deeper during the second diffusion than would be expected. It is as if the diffusing donor atoms push the diffusing acceptor atoms deeper than they would otherwise go. This is the emitter-dip effect.

The effect is most simply demonstrated by an experiment of the type described in Fig. 8.10. Fig. 8.10a represents the original slice after the first diffusion. An acceptor has been diffused in from the top surface. The surface is then covered with an oxide film, and a window is opened up for the emitter diffusion. The second diffusion then takes place, perhaps at about 1000°C. The donor diffuses through the window and the acceptor diffuses further into the silicon. Immediately below the emitter the acceptor penetration is greater than in the region remote from the emitter. This is shown in Fig. 8.10b, in which typical dimensions are given. The designation n^+ is given to the emitter doping merely to indicate that the emitter doping is greater than the base doping (otherwise the emitter would not be *n* type). The effect is only observed if the two *p–n* junctions are close together: the original base-collector junction should be 3 μm or less below the surface after the first diffusion. It is important only if the emitter is smaller than the collector, and if the diffusions are shallow. Such conditions are common in device manufacture, however, and the effect can make it difficult to achieve very narrow base widths.

Early experiments indicated that the effect was not caused by the oxide film but was due, rather, to an enhanced diffusion coefficient of the acceptor just below the emitter. It is usually assumed that, in silicon, group V donors and also group III acceptors diffuse by a substitutional mechanism. As noted above, such a mechanism gives a diffusion coefficient which is proportional to the concentration of vacancies in the lattice. The emitter-dip effect could, therefore, be due to an excess of vacancies being produced just below the emitter region. This was confirmed in a

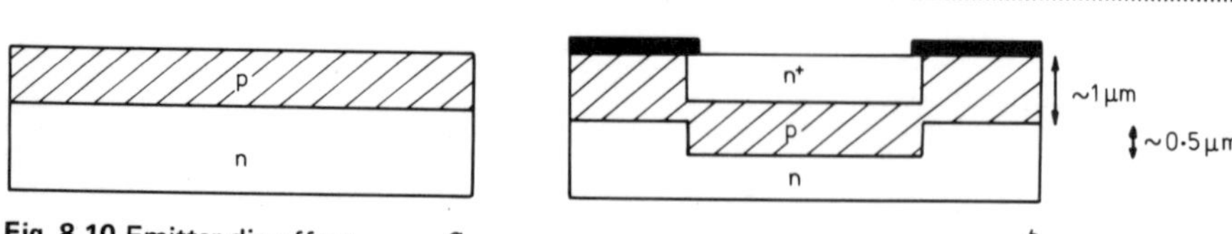

Fig. 8.10 Emitter-dip effect

striking way by some experiments carried out by Baruch, Constantin, Pfister and Saintesprit (1961)*. They used a beam of 275 keV protons to create vacancies just below the surface of a slice of silicon. The experiment was carried out as follows. They prepared a *p–n* junction in *n*-type silicon by diffusing in gallium to give a junction 6 μm below the surface. This gave a sample of the sort shown in Fig. 8.10a. They then irradiated the top surface of the sample with the proton beam. The width of the beam was restricted to 0·5 mm so that only a small part of the surface was bombarded with protons. The irradiation was carried out at about 1000° C, so that further diffusion of the gallium into the silicon was taking place all the while. At the end of the experiment, it was seen that a dip effect almost identical to the one already described had taken place. Below the irradiated strip, the *p–n* junction was deeper than elsewhere, giving a structure similar to that shown in Fig. 8.10b. The explanation for this result seems to be that where the proton beam struck, it produced vacancies just below the surface. They diffused into the interior of the crystal and brought about a local enhancement of the gallium diffusion coefficient, and a consequent deepening of the *p–n* junction.

Let us return to the conventional emitter-dip effect. The experiments of Baruch *et al.* suggest that the enhanced diffusion in the base is caused by extra vacancies flowing into the region just below the emitter. Taking this explanation as a working hypothesis, it becomes necessary to consider ways in which these extra vacancies might be produced. It was noted in Chapter 4 that vacancies can act as acceptors in silicon and that this leads to an *n*-type specimen containing more vacancies than a *p*-type specimen. This means that the emitter of an *n–p–n* structure contains more vacancies than the base, and it is tempting to think of this extra concentration as the source which produces the effect. There are difficulties in the way of this explanation, however. It must be pointed out that the extra vacancies cannot diffuse out of the emitter into the base in an isothermal experiment, even though there is a concentration gradient. Both the emitter and the base have the 'correct' equilibrium concentration of vacancies: it just happens that the two concentrations are different. There is no gradient in chemical potential and therefore no net movement of vacancies. The explanation, attractive though it may be at first sight, seems to have been finally eliminated by the work of Lawrence (1966)†, who observed the emitter-dip effect both in *p–n–p* structures and in $p^+ - p{-}n$. This second type is made by diffusing a high-concentration layer of an acceptor into a sample of the type shown in Fig. 8.10a. Neither of these types of specimen would have an excess of vacancies in the 'emitter'.

Lawrence also carried out a series of experiments in which pressure was applied to the top surface of a specimen while diffusion was taking place. Once again, he started with a specimen like that of Fig. 8.10a, formed by diffusing boron into

* Baruch, P., Constantin, C., Pfister, J. C., and Saintesprit, R., Discussion Faraday Soc., 1961, **31**, p. 76

† Lawrence, J. E., *J. Appl. Phys.*, 1966, **37**, p. 4106

n-type silicon. Pressure was applied to a small area of the surface using a silicon pin, and the specimen was taken up to diffusion temperature, so that the boron could continue to diffuse. The experiment was carried out for times ranging from a few minutes to several hours. At the end of an experiment, it was found that the *p–n* junction was deeper just below the pin than elsewhere, i.e. the effect of the pressure had been to increase the diffusion coefficient of the boron just below the pin. It was found that during the heat treatment, the silicon lattice beneath the pin deformed by a rapid dispersion of newly-created dislocations. The region containing the dislocations was quite well defined. It grew rapidly at the start of the experiment and stabilised after about ten minutes. When the progression of the dip effect was monitored against time, it was found that virtually all of the dip in the junction profile occurred during this first ten minutes. Lawrence repeated the experiment for gallium diffusing into *n*-type silicon and for antimony diffusing into *p*-type and obtained essentially the same result.

The fact that the dip effect in these experiments occurred while the pressure-induced dislocations were moving and stopped when a stable configuration was reached, suggest a reasonably straightforward interpretation for the experiments. The dislocations contained jogs which threw out vacancies when the dislocation lines moved. The vacancies caused a local increase in the diffusion coefficient, bringing about the dip effect. These ideas can be carried over to the original emitter-dip effect, replacing the pressure-induced dislocations with the diffusion-induced dislocations brought about by the emitter diffusion.

Only one experimental result casts a shadow over this rather happy state of affairs. Lawrence repeated the experiment, this time diffusing phosphorus into silicon. He observed a negative dip effect, i.e. the phosphorus immediately under the pin diffused less than that elsewhere. He also found that under certain circumstances it was possible to produce a negative emitter-dip effect in a *p–n–p* structure, when the *n*-type base was formed by phosphorus diffusion. The suggestion was made that the result was due to the phosphorus precipitating in the deformed region instead of diffusing in it. It is perhaps worth noting, also, that a dislocation that can emit vacancies by climbing in one direction can also absorb them by climbing in the other direction. It is not clear, however, why the direction of climb should be different for different impurities.

While the emitter-dip effect is not completely understood, it would appear, then, that it is probably associated with vacancy production by climbing dislocations. The effect has so far been studied only in silicon, but if the dislocation mechanism is correct, there seems to be no reason why it should not occur in other semiconductors.

Chapter 9

Appendixes

9.1 Diffusion coefficients and solubilities in silicon and germanium

The reader who has survived Chapters 3 and 5 will realise that the concepts of diffusion coefficient and solubility are not as straightforward as they appear at first sight. It would be pleasant, for instance, to think that a simple diffusion experiment would give both of these quantities directly. For copper diffusing in germanium, we might simply surround a germanium slice by an atmosphere of copper at 800°C and the shape of the profile would give a value for the diffusion coefficient at that temperature. Unfortunately, the experiment I have outlined is not unambiguous unless a liquid phase is formed in the system. Consider Fig. 3.6, which gives the Cu–Ge phase diagram. A horizontal line at 800°C cuts the solidus at a point which represents a solid solution of copper in germanium. This is the point which gives the solubility. The 800°C line cuts the liquidus at a point which gives the composition of the liquid which exists in equilibrium with that solid. This liquid composition will fix the equilibrium vapour pressure of copper above the solid and liquid phases.

Now suppose that in carrying out our diffusion experiment we impose a smaller copper pressure than this equilibrium pressure. In this case, there will be no liquid phase in the system. Because the germanium is surrounded by a lower vapour pressure than equilibrium, the surface concentration of the resulting diffusion profile will be less than the value given by the solidus line on the phase diagram. Thus the equilibrium solubility in such a system corresponds to the surface concentration of the diffusion profile only if all of the phases required by the phase diagram are present.

As if this were not enough, it happens very frequently that the shape of a diffusion profile does not fit neatly into one of the standard solutions outlined in Chapter 2, so a single value of diffusion coefficient cannot be fitted either. It has also been shown that the value of diffusion coefficient can depend on whether the semiconductor is n-type, p-type, or intrinsic (Fig. 4.3).

Having said all this, it is possible to compile a list of solubilities and diffusion coefficients of the more important dopants in silicon and germanium, on the understanding that the reader will look at them critically. Where he wants to be precise in his information, he should always refer to the original paper to ascertain the experimental conditions under which the results were obtained. Data for the 'typical' diffusion temperatures of 800°C for germanium and 1200°C for silicon is presented in Tables 9.1.1 and 9.1.2. Some of the values given in the tables are extrapolated from lower temperatures. Figs. 9.1.1 to 9.1.4 show how the diffusion coefficients for a number of these elements vary with temperature. Figs. 9.1.5 and 9.1.6 show the solid solubilities as functions of temperature. Each of the curves is equivalent to a solidus line on a phase diagram such as Fig. 3.6.

For diffusion in compound semiconductors the situation is more complicated and all of the warnings given above for silicon and germanium apply, several times over. If some dopant is diffused into a compound, then three elements are involved rather than two, and an extra degree of freedom is obtained for the phases that can exist (see the section on 'phase rule' in Chapter 3). In practical terms this means that the diffusion coefficient and solubility are no longer determined just by the temperature, so relatively simple data of the sort given above for silicon and germanium cannot be presented for compound semiconductors. The point is made

Table 9.1.1 Diffusion coefficients and solid solubilities in silicon at 1200°C

Element	D at 1200°C $cm^2\ s^{-1}$	Reference	Solid solubility at 1200°C cm^{-3}	Reference
Ag	$8{\cdot}0 \times 10^{-9}$	5	6×10^{15}	5
Al	$1{\cdot}5 \times 10^{-11}$	21	2×10^{19}	32
As	$2{\cdot}5 \times 10^{-13}$	19	2×10^{21}	32
Au	$1{\cdot}6 \times 10^{-6}$	30	8×10^{16}	32
B	$4{\cdot}5 \times 10^{-12}$	36	5×10^{20}	32
C	$3{\cdot}3 \times 10^{-11}$	23	—	
Cu	$1{\cdot}5 \times 10^{-5}$	4	1×10^{18}	32
Fe	$6{\cdot}6 \times 10^{-6}$	30	2×10^{16}	32
Ga	$3{\cdot}5 \times 10^{-12}$	12	4×10^{19}	32
In	$7{\cdot}3 \times 10^{-13}$	12	$>10^{19}$ at 1300°C	29
K	$2{\cdot}8 \times 10^{-6}$	31	7×10^{18}	31
Li	$1{\cdot}8 \times 10^{-5}$	26	6×10^{19}	32
Mn	$>2 \times 10^{-7}$	8	2×10^{16}	32
Na	$5{\cdot}7 \times 10^{-6}$	31	9×10^{18}	31
Ni	$3{\cdot}1 \times 10^{-8}$	6	6×10^{17}	37
P	$3{\cdot}0 \times 10^{-12}$	12	1×10^{21}	32
S	$2{\cdot}7 \times 10^{-8}$	9	2×10^{16}	32
Sb	3×10^{-13}	28	6×10^{19}	32
Si	$8{\cdot}4 \times 10^{-14}$	25	—	
Sn	$1{\cdot}0 \times 10^{-13}$	35	6×10^{19}	32
Zn	$\sim 5 \times 10^{-7}$	13	3×10^{16}	32

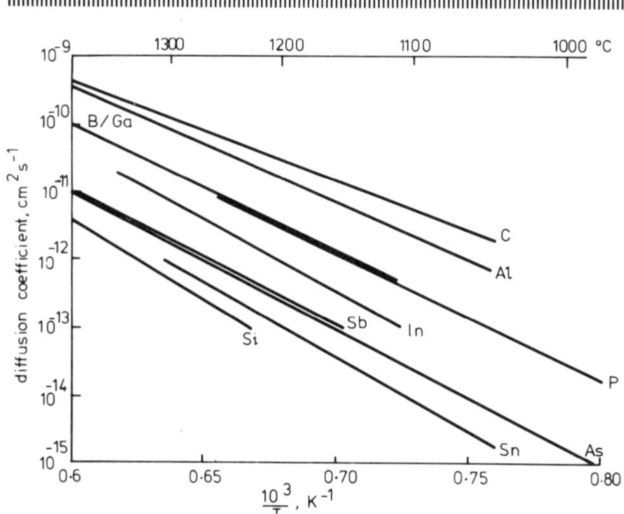

Fig. 9.1.1 Diffusion coefficients in silicon. The data is taken from the references given in Table 9.1.1. The lines for boron and gallium are sufficiently close to be represented by a single line

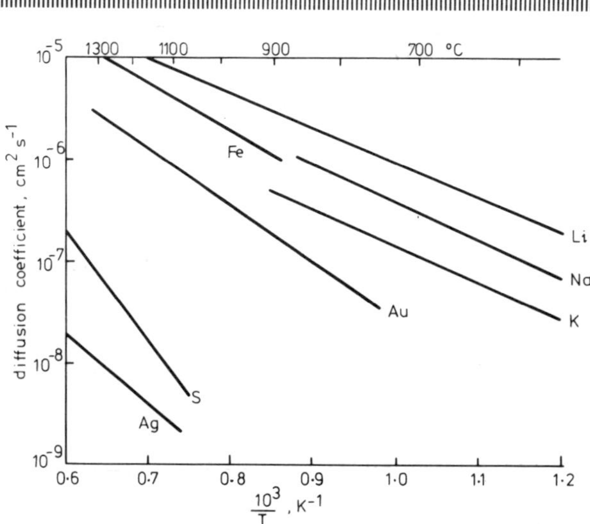

Fig. 9.1.2 Diffusion coefficients in silicon. The data is taken from the references given in Table 9.1.1

Table 9.1.2 Diffusion coefficients and solid solubilities in germanium at 800°C

Element	D at 800°C $cm^2\ s^{-1}$	Reference	Solid solubility at 800°C cm^{-3}	Reference
Ag	$8{\cdot}8 \times 10^{-7}$	7	4×10^{14}	32
Al	$1{\cdot}0 \times 10^{-13}$	20	4×10^{20}	32
As	$1{\cdot}0 \times 10^{-11}$	1	2×10^{20}	32
Au	$4{\cdot}0 \times 10^{-10}$	10	$\sim 10^{16}$	32
B	$7{\cdot}2 \times 10^{-15}$	20	$\sim 10^{18}$	14
Be	$1{\cdot}5 \times 10^{-12}$	2	1×10^{16}	2
Cd	$3{\cdot}6 \times 10^{-12}$	16	1×10^{18}	16
Cu	$2{\cdot}7 \times 10^{-5}$	3	2×10^{16}	32
Fe	$2{\cdot}3 \times 10^{-7}$	34	1×10^{15}	32
Ga	$8{\cdot}9 \times 10^{-14}$	11	4×10^{20}	32
Ge	$8{\cdot}0 \times 10^{-14}$	17	—	—
In	$9{\cdot}5 \times 10^{-13}$	24	4×10^{18}	32
Li	$1{\cdot}9 \times 10^{-5}$	27	7×10^{18}	32
Ni	$4{\cdot}4 \times 10^{-5}$	18	3×10^{15}	32
P	$8{\cdot}0 \times 10^{-12}$	11	—	—
S	$\sim 10^{-9}$ at 920°C	33	—	—
Sb	$3{\cdot}1 \times 10^{-11}$	22	1×10^{19}	32
Se	$\sim 10^{-10}$ at 920°C	33	—	—
Sn	$2{\cdot}0 \times 10^{-11}$	3	$2{\cdot}5 \times 10^{20}$	32
Te	$2{\cdot}1 \times 10^{-11}$	15	3×10^{18}	15
Zn	$1{\cdot}0 \times 10^{-12}$	11	$2{\cdot}5 \times 10^{18}$	32

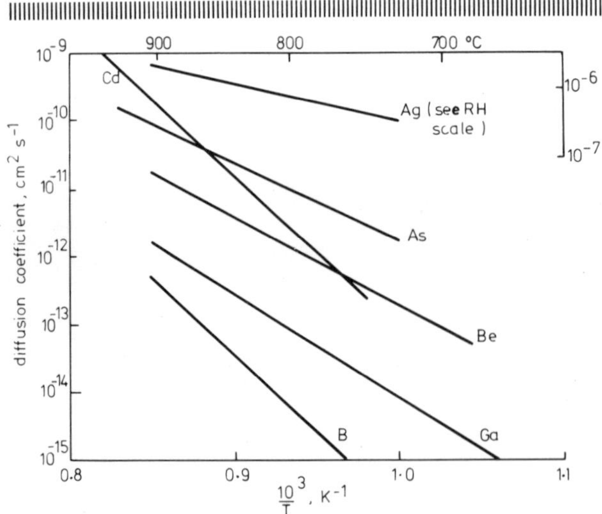

Fig. 9.1.3 Diffusion coefficients in germanium. The data is taken from the references given in Table 9.1.2. Note the separate scale for silver

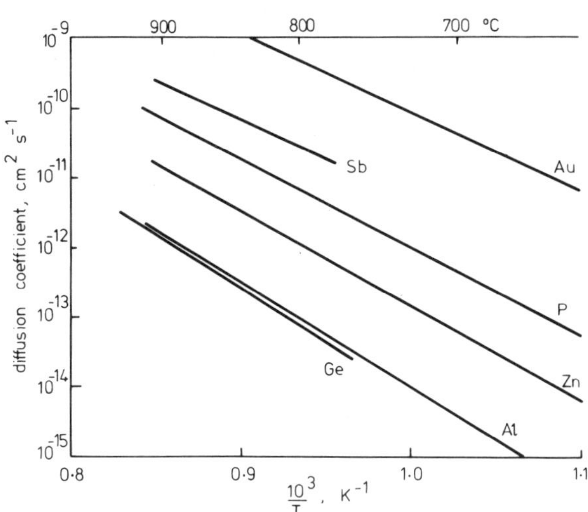

Fig. 9.1.4 Diffusion coefficients in germanium. The data is taken from the references given in Table 9.1.2

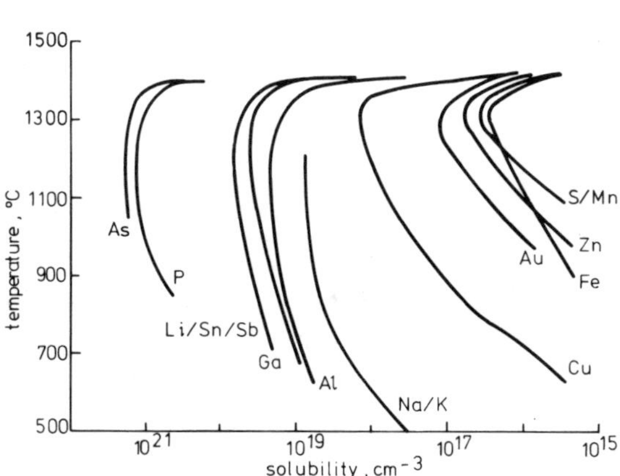

Fig. 9.1.5 Solid solubilities in silicon. The data is taken from the references of Table 9.1.1. In some cases the curves for two different elements are so close together that they can be described by a single curve

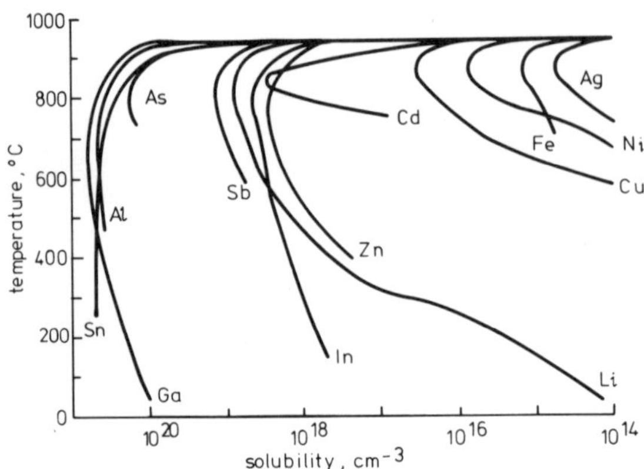

Fig. 9.1.6 Solid solubilities in germanium. The data is taken from the references of Table 9.1.2

eloquently in Fig. 6.9 and Fig. 6.10, in which the diffusion coefficient of zinc in GaAs is shown to vary by a factor of 10^5 at a single temperature. When interpreting experimental data for diffusion coefficient and solubility in compounds, therefore, it is necessary to know the conditions under which they were determined, and the reader is urged to consult the original literature. A comprehensive list of references for diffusion in a wide range of semiconductors has been compiled by B. L. Sharma in 'Diffusion in semiconductors', published by TransTech Publications. It is complete up to 1970.

9.1.1 References

1 ALBERS, W.: *Solid State Electron.*, 1961, **2**, p. 85
2 BELYAEV, Y. I., and ZHIDKOV, V. A.: *Sov. Phys. – Solid State,* 1961, **3**, p. 133
3 BOLTAKS, B. I.: 'Diffusion in Semiconductors', Infosearch, London, 1963, p. 162
4 BOLTAKS, B. I., and SOSINOV, I. I.: *Zh. Tekh. Fiz.*, 1958, **28**, 3
5 BOLTAKS, B. I., and SHIH-YIN, H.: *Sov. Phys. – Solid State*, 1961, **2**, p. 2383
6 BONZEL, H. P.: *Phys. Status Solid,* 1967, **20**, p. 493
7 BUGAI, A. A., KOSENKO, V. E., and MISELYUK, E. G.: *Sov. Phys. Tech. Phys.*, 1957, **2**, p. 1553
8 CARLSON, R. O.: *Phys. Rev.*, 1956, **104**, p. 937
9 CARLSON, R. O., HALL, R. N., and PELL, E. M.: *J. Phys. & Chem. Solids,* 1959, **8**, p. 81
10 DUNLAP, W. C.: *Phys. Rev.,* 1955, **97**, p. 614
11 DUNLAP, W. C.: *Phys. Rev.,* 1954, **94**, p. 1531
12 FULLER, C. S., and DITZENBERGER, J. A.: *J. Appl. Phys.*, 1956, **27**, p. 544
13 FULLER, C. S., and MORIN, F. J.: *Phys. Rev.,* 1957, **105**, p. 379

14 GLAZOV, V. M., and ZEMSKOV, V. S.: Fiz.-Khim. Osnovi Lagirovaniya Poluprovodnikov, Isdatelstvo, Moscow, 1968 (Quoted in B. L. Sharma, 'Diffusion in Semiconductors', Trans. Tech. Publ., Clausthal-Zellerfeld, 1970)

15 IGNATKOV, V. D., and KOSENKO, V. E.: *Sov. Phys. – Solid State*, 1962, **4**, p. 1193

16 KOSENKO, V. E.: *Sov. Phys. – Solid State,* 1960, **1**, p. 1481

17 LETAW, H., PORTNOY, W. M., and SLIFKIN, L. M.: *Phys. Rev.*, 1956, **102**, p. 636

18 VAN DER MAESSEN, F., and BRENKMAN, J. A.: Philips Res. Repts., 1954, **9**, p. 225

19 MASTERS, B. J., and FAIRFIELD, J. M.: *J. Appl. Phys.*, 1969, **40**, p. 2390

20 MEER, W., and POMERRENIG, D.: *Zeit. Angew. Phys.*, 1967, **23**, p. 369

21 MILLER, R. C., and SAVAGE, A.: *J. Appl. Phys.*, 1956, **27**, p. 1430

22 MILLER, R. C., and SMITS, F. M.: *Phys. Rev.*, 1957, **107**, p. 65

23 NEWMAN, R. C., and WAKEFIELD, J.: *J. Phys. & Chem. Solids,* 1961, **19**, p. 230

24 PANTALEEV, V. A.: *Sov. Phys. – Solid State,* 1965, **7**, p. 734

25 PEART, R. F.: *Phys. Status Solidi*, 1966, **15**, p. K119

26 PELL, E. M.: *Phys. Rev.*, 1960, **119**, p. 1014

27 PRATT, B., and FRIEDMAN, F.: *J. Appl. Phys.*, 1966, **37**, p. 1893

28 ROHAN, J. J., PICKERING, N. E., and KENNEDY, J.: *J. Electrochem. Soc.*, 1959, **106**, p. 705

29 SMITH, A. M.: 'Fundamentals of Silicon Integrated Device Technology' Ed. R. M. Burger and R. P. Donovan, Prentice-Hall, N.J., 1967, p. 255

30 STRUTHERS, J. D.: *J. Appl. Phys.*, 1956, **27**, p. 1560. See also *J. Appl. Phys.*, 1957, **28**, p. 516 for a correction to the first paper

31 SVOB, L.: *Solid State Electron.*, 1967, **10**, p. 991

32 TRUMBORE, F. A.: *Bell Syst. Tech. J.*, 1960, **39**, p. 205

33 TYLER, W. W.: *J. Phys. & Chem. Solids,* 1959, **8**, p. 59

34 WEI, L. Y.: *J. Phys. & Chem. Solids,* 1961, **18**, p. 162

35 YEH, T. H., HU, S. M., and KASTL, R. H.: *J. Appl. Phys.*, 1968, **39**, p. 4266

36 YOMAGUCHI, J., HORIUCHI, S., MATSAMURA, K., and OEJINO, Y.: *J. Phys. Soc. (Japan)*, 1960, **15**, p. 1541

37 YOSHIDA, M., and FURUSHO, K.: *J. Appl. Phys. (Japan)*, 1964, **3**, p. 521

9.2 Relationship between average conductivity and surface concentration for a diffused layer

In principle, if a diffusion brings about a *p–n* junction in a semiconductor, the surface concentration of the diffusion profile is specified by the following:

(*a*) the mathematical form of the diffusion profile
(*b*) data showing how electron and hole mobilities vary with doping of donors and acceptors, respectively
(*c*) average conductivity of the diffused layer

Suppose we carry out a diffusion experiment, introducing donors into a slice of *p*-type semiconductor. The original background doping is C_B (acceptors). The surface concentration of the profile is C_0 (donors). Assume $C_0 > C_B$ so that a p–n junction is formed. If the junction is at depth x_0 below the surface, we have the boundary conditions

$$C = C_0 \text{ at } x = 0 \tag{9.2.1}$$

$$C = C_B \text{ at } x = x_0 \tag{9.2.2}$$

The data mentioned in (a–c), above, is obtained as follows:

(a) The mathematical form, i.e. error function complement, Gaussian etc. may be given in the diffusion literature. Alternatively we might make a guess and assume some simple form for the profile

(b) This information is available for some semiconductors. Fig. 9.2.1 shows how the mobilities of n- and p-type silicon, germanium and GaAs vary with doping. The data is taken from Sze and Irvin (1968)

(c) This is determined experimentally using the four point probe and a determination of junction depth, x_0. The average conductivity of the diffused layer $\bar{\sigma}$ is determined using eqn. 7.15, substituting x_0 for w, i.e.

$$\bar{\sigma} = \frac{1}{4{\cdot}5x_0} \cdot \frac{I}{V} \tag{9.2.3}$$

where the reader is reminded that if the spacing between the probes is not greater than $2x_0$, the figure of 4.5 has to be modified (see Chapter 7). It is assumed here, as in Chapter 7, that the p–n junction provides an open circuit so that the measured conductivity is just that of the diffused layer.

Now consider how this information can be used to calculate the surface concentration. The average conductivity of the layer is also given by

$$\bar{\sigma} = \frac{e}{x_0} \int_0^{x_0} B(n)\,[C(x) - C_B]\,dx \tag{9.2.4}$$

$\bar{\sigma}$ and x_0 are known from the experimental measurements. C_B is a property of the starting material which can easily be measured if it is not known already. (It is assumed here for simplicity that all donors and acceptors are ionised. It is quite possible to add Fermi-Dirac statistics to the above expression to allow for the fact that some will not be ionised.) $B(n)$, the electron mobility, is given in Fig. 9.2.1. Let us assume that $C(x)$ is of the form $C_0 f(x)$, where $f(x)$ is some known function which varies between one and zero. This leaves C_0 as the only unknown in eqn. 9.2.4. The equation can therefore be solved for C_0 subject to the boundary conditions eqns. 9.2.1 and 9.2.2. This is done using a computer and typical solutions are given in Fig. 9.2.2 for silicon and Fig. 9.2.3 for germanium. The solutions are for the two main types of diffusion profile; error function complement and Gaussian. The data is taken from Irvin (1962) and Cuttriss (1961): the original articles contain many

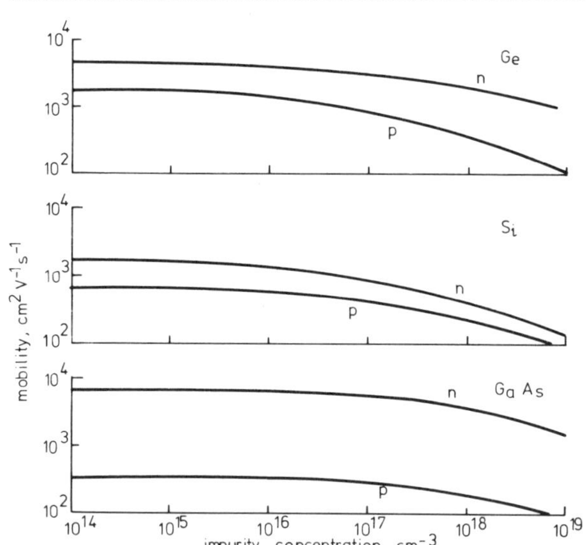

Fig. 9.2.1 Electron and hole mobilities in silicon, germanium and GaAs at 300 K

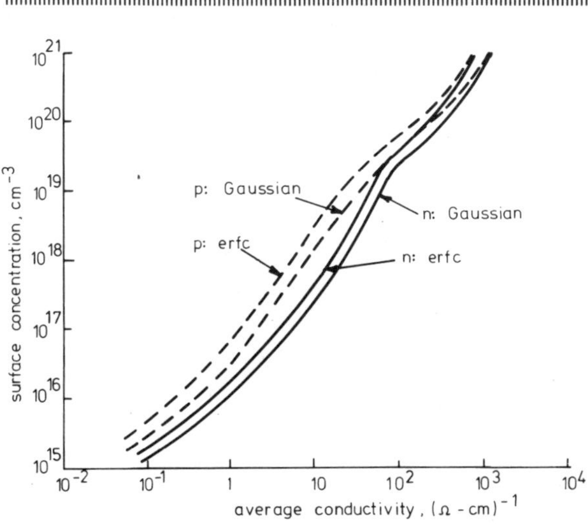

Fig. 9.2.2 Surface concentration as a function of average conductivity in silicon at 300 K. Curves are plotted for error function complement and Gaussian profiles in both *n*- and *p*-type material. Background doping is 10^{15} cm^{-3} in all cases

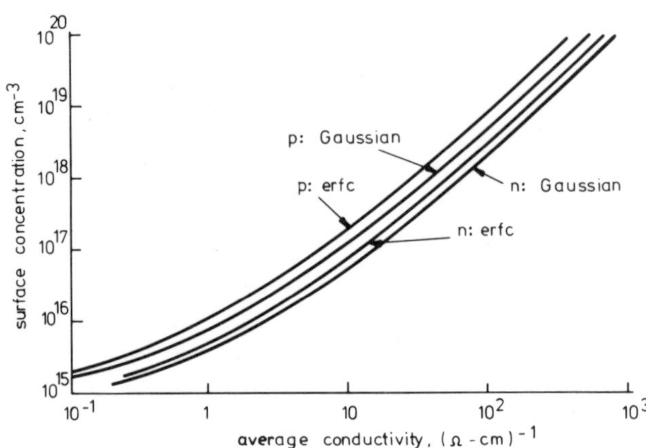

Fig. 9.2.3 Surface concentration as a function of average conductivity in germanium at 300 K. Curves are plotted for error function complement and Gaussian profiles in both *n*- and *p*-type material. Background doping is 10^{15} cm^{-3} in all cases

more graphs, plotting $\bar{\sigma}$ against C_0 for a variety of background concentrations C_B and for profiles following different mathematical functions.

Figs. 9.2.2 and 9.2.3 provide a simple means of determining surface concentrations in silicon and germanium. The procedure is to measure the sheet conductivity and the junction depth, and hence calculate $\bar{\sigma}$. The surface concentration can then be read off one of the graphs. Implicit in this technique, however, is the assumption that the diffusion profile follows one of the 'standard' forms. It is to be hoped that the reader will look rather hard at this assumption when using the technique. For this reason, $\bar{\sigma}$ against C_0 curves are not presented for GaAs, even though the data of Fig. 9.2.1 makes it possible to calculate them. Generally speaking, the profiles observed in GaAs are neither standard nor simple.

9.2.1 References

CUTTRISS, D. B.: *Bell Syst. Tech. J.*, 1961, **40**, p. 509

IRVIN, J. C.: *Bell Syst. Tech. J.*, 1962, **41**, p. 387

SZE, S. M., and IRVIN, J. C.: *Solid State Electron.*, 1968, **11**, p. 599

Index